NORTH AMERICAN FIELD GUIDES

TRACKS AND SCAT

Kristin J. Russo

Field Guides

An Imprint of Abdo Reference | abdobooks.com

CONTENTS

WHAT ARE TRACKS AND SCAT?

Many kinds of animals live in North America. Sometimes people encounter these animals and observe how they behave. Hikers might see a bobcat on a mountain trail. A homeowner might find a raccoon or a group of crows rummaging through a garbage can. A birdwatcher might see a hawk circling overhead. Some wild animals are harder to observe. Owls and bats hunt at night and are hard to see in the dark. Bighorn sheep live in harsh mountain conditions where people rarely go.

All animals—even secretive, hard-to-observe ones—leave signs. Chunky white pellets filled with animal bones and fur along a woodland trail reveal that an owl has recently had a big meal. Nibbled pine branches low to the ground suggest that a snowshoe hare has hopped by. Animals also leave footprints and fecal droppings, called scat. Tracks and scat tell stories about where they've been, where they're going, and what they like to eat.

WHAT DO TRACKS AND SCAT LOOK LIKE?

Tracks are imprints that animals make in the sand, snow, soil, or mud. Usually, tracks are made by the animal's feet, but some birds, such as owls and hawks, leave wing tracks in the snow where they have pounced on their prey. Sidewinder snakes leave swooping, S-shaped curves in the sand. Polar bears leave enormous, claw-footed prints in the snow. And duck feet make webbed triangles in mud near ponds and rivers.

People study animal tracks to learn about the gait, or the way the animals walk. Some animals, such as ungulates that have hooves, are diagonal walkers. They move their right front leg at the same time as their back left leg, and vice versa. Bounders such as weasels and badgers make small jumps. Pacers such as porcupines and skunks often move in a slow shuffle.

Not all animals leave footprints, but all animals poop. Scientists study animal scat to learn what animals eat and what their digestive system is like. Tracks and scat cannot always be studied together. Some animals may leave tracks on land, but they defecate in the water, so their scat cannot be observed easily. Other animals, such as bats, leave abundant scat but not many tracks.

WHAT ROLE DO TRACKS AND SCAT PLAY?

Observing tracks and scat gives scientists important information about how animals live, hunt, and socialize. Scientists study tracks to determine if animals have enough to eat or if they need to travel to find food. Scat can reveal what types of animals live in an area and if they are thriving or struggling due to lack of food or water.

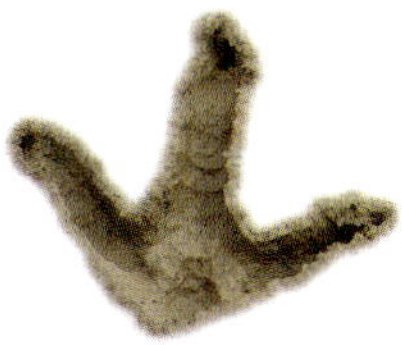

HOW TO USE THIS BOOK

Tab shows the animal's category.

The featured animal's common name appears here.

BIRDS

AMERICAN CROW

(CORVUS BRACHYRHYNCHOS)

American crows are abundant in North America. Scientists who study these black-feathered birds report that they are very smart. They bend twigs to make hooks and use them to spear and pry food from tiny spaces in tree bark. Instead of dropping their hooks while they eat, these clever birds place them in a small tree hole or hold them with their foot. When … to feed on the ground, they leave tracks near … have four toes with claws on each … ard and one faces backward.

***Fun Facts* give interesting information related to the tracks or scat.**

FUN FACT

American crows live and travel in large groups called murders. Sometimes up to five generations of the same family live in the group.

Images show the animal and tracks or scat.

HOW TO SPOT

Track Size: 3 inches (7.6 cm) long
Range: United States and Canada
Habitat: Fields, orchards, savannas, suburbs
Diet: Small animals, berries, insects, bird eggs, and carrion

10

AMERICAN FLAMINGO

(PHOENICOPTERUS RUBER)

The animal's scientific name appears here.

Adult American flamingos are bright pink.The type of food they eat—brine shrimp and aquatic plants—creates a chemical reaction in their body that turns their feathers pink. Flamingos often stand on one foot. They tend to stand on one foot more often in the water than on land, birds' three-toed, webbed footprints can be found in sandy soil. In fact, these birds live in large grou where there is one flamingo track, there are usua of others.

The paragraph gives information about the tracks or scat.

HOW TO SPOT

Track Size: About 3.5 inches (8.9 cm) wide and 2.9 inches (7.4 cm) long

Range: Southeaste United States; Yuc Peninsula in Mexic and Caribbean isla

Habitat: Saltwate shallows

Diet: Algae, small seeds, tiny crusta and fly larvae

***How to Spot* features give information about the animal's track or scat size, range in North America, habitat, and diet.**

EATING UPSIDE DOWN

The American flamingo is a filter feeder. It eats with its head upside down and its bill facing toward its feet. As it sweeps its head back and forth, it uses its tongue to pump water in and out of its bill. Small, comb-like plates filter the water out and trap the food in

11

Sidebars provide additional information about the topic.

ARIZONA BARK SCORPION

(CENTRUROIDES SCULPTURATUS)

The Arizona bark scorpion's brown body is separated into two segments. It has eight legs plus two front legs that look like small lobster claws. These front legs, called pedipalps, grab prey and help the scorpion feel its surroundings. This scorpion has a tail that curls over its body with a stinger at the end. A person stung by an Arizona bark scorpion should seek medical help right away. It is one of the deadliest scorpions. Though this scorpion is small and light, it leaves small footprint tracks as it ambles along in the desert sand.

HOW TO SPOT

Track Size: A set of 8 tracks is about 1.5 inches (3.8 cm) long

Range: Mexico and southern United States

Habitat: Tree bark, logs, and rock crevices

Diet: Insects, spiders, and other scorpions

ATLANTIC HORSESHOE CRAB

(LIMULUS POLYPHEMUS)

The horseshoe crab is a unique animal. It gets its name from its semicircular shape—like a horseshoe. If this odd-looking creature wore shoes, it would need 12! It has 12 legs but uses only 10 of them to walk. Because this animal lacks a jaw and cannot chew, it uses its front two legs to crush food and move it into its mouth. When a horseshoe crab tips over, it uses its straight, sharp tail to dig into the sand and turn itself upright. A horseshoe crab leaves tracks in the sand that show the outline of its hard shell. Its pointy tail leaves a trail behind.

HOW TO SPOT

Track Size: 12 inches (30.5 cm) wide

Range: Atlantic and Gulf coasts of the United States and Yucatán Peninsula in Mexico

Habitat: Sandy beaches

Diet: Worms, algae, clams, and other small prey

WHAT IS AN ARTHROPOD?

The term *arthropod* comes from a Latin word that means "jointed feet." A close look at an arthropod will show that its feet, legs, and body are separated into segments. An arthropod's body is covered with a hard shell, or exoskeleton. A surprising group of animals and insects are arthropods, including lobsters, crabs, spiders, mites, insects, centipedes, and millipedes.

AMERICAN CROW

(CORVUS BRACHYRHYNCHOS)

American crows are abundant in North America. Scientists who study these black-feathered birds report that they are very smart. They bend twigs to make hooks and use them to spear and pry food from tiny spaces in tree bark. Instead of dropping their hooks while they eat, these clever birds place them in a small tree hole or hold them with their foot. When crows gather to feed on the ground, they leave tracks near their feeding sites. Crows have four toes with claws on each foot. Three toes face forward and one faces backward.

FUN FACT

American crows live and travel in large groups called murders. Sometimes up to five generations of the same family live in the group.

HOW TO SPOT

Track Size: 3 inches (7.6 cm) long
Range: United States and Canada
Habitat: Fields, orchards, savannas, suburbs
Diet: Small animals, berries, insects, bird eggs, and carrion

AMERICAN FLAMINGO

(PHOENICOPTERUS RUBER)

Adult American flamingos are bright pink. The type of food they eat—brine shrimp and aquatic plants—creates a chemical reaction in their body that turns their feathers pink. Flamingos often stand on one foot. They tend to stand on one foot more often in the water than on land, so the birds' three-toed, webbed footprints can be found together in sandy soil. In fact, these birds live in large groups, so where there is one flamingo track, there are usually plenty of others.

HOW TO SPOT

Track Size: About 3.5 inches (8.9 cm) wide and 2.9 inches (7.4 cm) long

Range: Southeastern United States, Yucatán Peninsula in Mexico, and Caribbean islands

Habitat: Saltwater shallows

Diet: Algae, small seeds, tiny crustaceans, and fly larvae

EATING UPSIDE DOWN

The American flamingo is a filter feeder. It eats with its head upside down and its bill facing toward its feet. As it sweeps its head back and forth, it uses its tongue to pump water in and out of its bill. Small, comb-like plates filter the water out and trap the food inside.

AMERICAN HERRING GULL

(LARUS SMITHSONIANUS)

The American herring gull is one of the most common seagulls in North America. These gulls have clever ways to make eating easier. To crack open hard mollusk shells, they hover above jetties and drop their prey onto the rocks below. Sometimes they hold bread in their bills to "feed" fish and then snatch their prey when the fish come to take a bite. Like most seabirds, American herring gulls have webbed feet to help them swim. When they walk on the beach, three front toes show up in their tracks, but a smaller back claw does not usually make a mark in the sand.

HOW TO SPOT

Track Size: 3 inches (7.6 cm) long

Range: United States, Canada, northern Mexico, and Caribbean islands

Habitat: Sandy coasts, lakes, rivers, and landfills

Diet: Fish, crustaceans, mollusks, sea urchins, marine worms, birds, eggs, insects, garbage, and carrion

FOOT STOMPERS

American herring gulls stomp their feet on the ground to mimic the vibrations of a heavy rainstorm. Worms and other prey rise to the surface to avoid getting soaked underground. Then they become a tasty meal for the clever gulls.

AMERICAN WHITE PELICAN

(PELECANUS ERYTHRORHYNCHOS)

American white pelicans are hefty waterbirds. They have large, roomy bills that they use to scoop fish out of the water. These birds often hunt in groups and build nests near one another. In fact, during migration and breeding season, American white pelicans gather in colonies, or large groups, of several thousand! Their tracks show that they have four toes that all face forward and are fully webbed. Their webbed feet help them swim, but they are clumsy walkers.

HOW TO SPOT

Track Size: About 6.2 to 7 inches (15.7 to 17.8 cm) long and 3.5 to 4.5 inches (8.9 to 11.4 cm) wide

Range: Central and southern Canada, northwestern and southeastern United States, and Mexico

Habitat: Lakes, marshes, and salt bays

Diet: Fish, crustaceans, and amphibians

BALD EAGLE

(HALIAEETUS LEUCOCEPHALUS)

Bald eagles are large raptors. They are graceful in flight but awkward on land. They have thick, stocky legs and long talons that make them clumsy walkers. Eagles hunt fish and seabirds. Their tracks are most often found in soft, sandy soil along riverbeds where their prey is most abundant. As they walk or hop, they leave tracks with three toes facing forward and one toe facing backward. Their kill tracks include their broad wingspan and often the footprints of their fleeing prey.

HOW TO SPOT

Track Size: 6 to 7.5 inches (15.2 to 19.1 cm) long
Range: United States, Canada, and northern Mexico
Habitat: Marshes, rivers, and streams
Diet: Fish, rodents, mammals, and carrion

BACK FROM THE BRINK

In the mid-1900s, the bald eagle almost went extinct. By 1963, only 471 pairs of bald eagles were left on Earth. Since then, their population has rebounded. They are no longer considered an endangered species.

GREATER ROADRUNNER

(GEOCOCCYX CALIFORNIANUS)

The greater roadrunner is the largest type of cuckoo bird in North America. It has wings and can fly, but because most of its prey is on the ground, it usually walks or runs. With long legs and long tail feathers that help it balance, it can run up to 15 miles per hour (24.1 kmh). It can go even faster in short spurts to chase its prey. Roadrunners have four toes per foot. Two toes on each foot face forward, and two face backward. This helps them grip their slippery prey. Their footprints leave zigzagging X marks in the desert sand.

FUN FACT

Greater roadrunners survive and even thrive in the harsh desert environment. They absorb the moisture their body needs through the food they eat, so they do not need to drink water.

HOW TO SPOT

Track Size: 3 inches (7.6 cm) long

Range: Southwestern United States and northern Mexico

Habitat: Deserts, chaparral, grasslands, woodlands, and agricultural areas

Diet: Small reptiles and mammals, scorpions, spiders, ground nesting birds, insects, fruits, and seeds

GREAT HORNED OWL

(BUBO VIRGINIANUS)

Great horned owls have a broad wingspan. Sometimes their wing prints are captured in snow or sand when they swoop down to grab a fleeing mouse or squirrel. These owls have brown and white feathers that provide camouflage for hiding in trees. They are active at night and are rarely seen in the day. Their unique tracks can be the only evidence that the large owl is nearby. Great horned owls have four toes that can face both backward and forward.This flexibility helps them grasp fleeing prey or cling to a perch.

HOW TO SPOT

Track Size: 4 inches (10.1 cm) long

Range: United States, Canada, and northern Mexico

Habitat: Forests, canyons, and clearings

Diet: Small mammals and other birds such as ducks and hawks

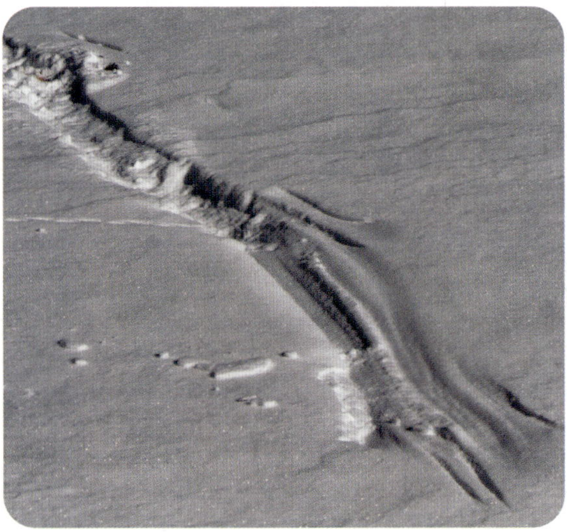

MALLARD DUCK

(ANAS PLATYRHYNCHOS)

Mallard ducks are colorful waterbirds. To eat, they tip their bodies face down into the water with their tails in the air. This is called dabbling. When they dabble, their bills can reach the plants and small animals at the muddy bottoms of the ponds, lakes, streams, and swamps where they live. Because they are waterbirds, their feet are webbed. Three front toes face forward. A rear toe that faces backward, the hallux, is small and doesn't often show up in their prints.

HOW TO SPOT

Track Size: 3.2 inches (8.1 cm) long and 3.7 inches (9.4 cm) wide

Range: Canada, United States and Hawaii, northern Mexico, and Caribbean islands

Habitat: Ponds, lakes, streams, and swamps

Diet: Aquatic plants, insects, small fish, tadpoles, freshwater snails, fish eggs, and frogs

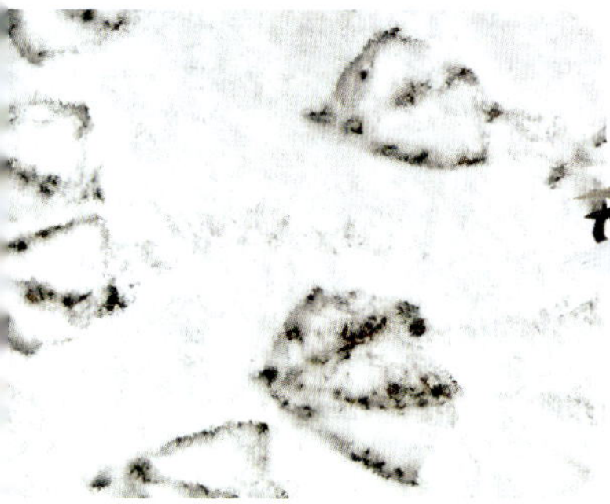

COLORFUL COURTSHIP

When mallards are choosing partners in the winter, their yellow bills and feet turn bright orange or red. In summer, when partners are nesting and raising their young, their bills and feet turn back to dull yellow. Male mallards change color, and their vibrant feathers become drab. This change in color helps them hide from predators in their nests.

RUFFED GROUSE *(BONASA UMBELLUS)*

The ruffed grouse walks more than it flies. These birds fly short distances from tree to tree but usually hide in low bushes. Its tracks show that its feet move in a straight line, one foot in front of the other. Their tracks reveal that grouses have three forward-facing toes and one smaller toe that points backward. When ruffed grouses walk, their footprints are about 3 inches (7.6 cm) apart. When the birds run, the prints can be up to 8 to 10 inches (20.3 to 25.4 cm) apart.

FUN FACT

People walking past a hiding ruffed grouse might get a big surprise. Suddenly, the silent grouse will make a loud flutter of noise as it moves away in a flurry of leaves or snow.

HOW TO SPOT

Track Size: 2 to 2.7 inches (5.1 to 6.9 cm) long

Range: Canada and the Appalachian Mountains in the United States

Habitat: Aspen stands and brush

Diet: Seeds, buds, fruits, berries, insects, and acorns

WILD TURKEY

(MELEAGRIS GALLOPAVO)

Wild turkeys are large, plump birds. Adult male turkeys are called gobblers. To attract female mates and discourage other males, gobblers puff out their body feathers, flare their tails, and holler a loud gobbling sound. Thanks to these loud, vibrant displays, it's not hard to find wild turkeys during mating season. But during their quieter months, finding their tracks might be the only way to tell if they are nearby. Turkeys have three toes facing forward and a hallux. They have claws on their thick, strong toes that help them dig for insects, roots, and tubers.

HOW TO SPOT

Track Size: About 4.5 inches (11.4 cm) long

Range: Southern Canada, most of the United States, and northern and central Mexico

Habitat: Forests, woodlands, and rural suburban areas

Diet: Nuts, berries, insects, and snails

FUN FACT

Although they do not have webbed feet, wild turkeys are capable swimmers. They live in forests that sometimes border streams and rivers.

WILLOW PTARMIGAN

(LAGOPUS LAGOPUS)

The willow ptarmigan is a North American bird that changes its plumage, or the color of its feathers, from summer to winter. In summer, its feathers are a mix of reds and browns that serve as camouflage, helping it hide among the mountain shrubs and trees in its habitat. In winter, this Arctic bird turns white, helping it hide almost completely in its snowy surroundings. Willow ptarmigans have heavily feathered feet that act as snowshoes when they walk, spreading out and keeping their feet from sinking.

HOW TO SPOT

Track Size: 2.5 inches (6.4 cm) long
Range: Northern Canada and Alaska in the United States
Habitat: Thickets, forests, tundra, and alpine meadows
Diet: Leaves and shoots of willow and birch trees, berries, seeds, and insects

SNOWBANK SNOOZING

Willow ptarmigans try not to leave any tracks, especially when they want to rest and hide from predators. Instead of walking or hopping, they fly into snowbanks and nestle in the snow to sleep. By diving in flight, they don't leave tracks for predators to follow.

WOOD DUCK *(AIX SPONSA)*

The wood duck is recognizable by its vibrantly colored feathers and the curved crest on its head. Wood ducks lay two separate groups of eggs. Instead of building a second nest, they may lay their eggs in another wood duck's nest. This is called nest dumping. The wood duck gets its name from its habitat. It lives in the woods and nests in tree hollows. Wood ducks have webbed toes like other ducks, but they also have strong claws. This helps them grip bark and perch high up on tree branches.

HOW TO SPOT

Track Size: About 3 inches (7.6 cm) long

Range: Southern Canada, most of the United States, northern and central Mexico, and Cuba

Habitat: Wooded swamps

Diet: Seeds, wetland plants, grasses, acorns, wild rice, pondweeds, insects, and invertebrates

AMERICAN BEAVER

(CASTOR CANADENSIS)

The American beaver needs to live in deeper pools of water than those usually found in wooded wetlands, so it builds itself a dam. It uses its strong teeth to chomp down trees, piling them in a spot that slows the flow of water and creates a pool. American beavers leave triangle-shaped tracks that are easy to spot in the mud around their dams and dens. Their five fingers show up in their webbed front tracks. Their tail often shows up in their tracks as well. It is heavy and will drag between the paw prints.

HOW TO SPOT

Track Size: Front tracks 2 to 3 inches (5.1 to 7.6 cm) long; hind tracks 6 to 7 inches (15.2 to 17.8 cm) long

Range: North America except southwestern United States

Habitat: Ponds, lakes, rivers, and streams

Diet: Leaves, woody stems, and aquatic plants

FUN FACT

The American beaver is the largest rodent in North America. The North American porcupine is the second-largest rodent on the continent.

AMERICAN MINK

(NEOVISON VISON)

The American mink is a weasel that spends time in the water and on land. It has dark brown or black fur with a small white patch on its chin and sometimes a white spot on its chest. Like most weasels, it has a long body and short legs. Minks hunt small animals, and they are strong predators. Sometimes they kill more than they need to eat, and they store food in their dens. Their five-toed feet are partially webbed, which helps them swim. Each toe has a claw that minks use to hunt and climb trees.

HOW TO SPOT

Track Size: 1.3 inches (3.3 cm) wide
Range: Canada and the United States, except Arizona
Habitat: Forests with a water source
Diet: Muskrats, rabbits, birds, frogs, fish, crayfish, and insects

FUN FACT

Minks purr like cats when they are happy. They hiss when they feel threatened.

ARCTIC FOX *(VULPES LAGOPUS)*

The Arctic fox lives in one of the harshest environments on Earth—the extreme cold of the Arctic north. Its body is shorter and stouter than other types of foxes. This helps it retain body heat. And its already dense fur grows even thicker in the winter. Arctic foxes have such heavy winter coats that they do not require winter hibernation, or deep rest, like many other animals do. These foxes hunt year-round, crossing the tundra in search of prey. Often they follow polar bears, hoping to scavenge leftovers from a bear kill. In the snow, they leave four-toed, oval- or diamond-shaped tracks.

HOW TO SPOT

Track Size: About 1.5 inches (3.8 cm) wide and 2 inches (5.1 cm) long

Range: Northern Canada and northern Alaska in the United States

Habitat: Tundra

Diet: Eggs, young birds, small rodents, ringed seal pups, grouse, berries, seaweed, insects, and frozen carrion

FUN FACT

The Arctic fox is the only wild dog that changes the color of its coat. Its fur is dark gray to bluish brown in the summer and white in the winter.

BIGHORN SHEEP *(OVIS CANADENSIS)*

The bighorn sheep is known for its large, curled horns. On males, called rams, the horns are a symbol of status. Males also use their horns as weapons when they fight. Despite having large, stocky bodies and stiff ungulate feet, these sheep can run and climb nimbly on steep mountain slopes and slim ridges. Bighorn sheep hoof prints resemble those of deer, but with some differences. Bighorn sheep hooves have straighter edges and are blockier and less pointed than those of deer. They often do not have the same heart shape seen in deer tracks.

HOW TO SPOT

Track Size: Front tracks 2.1 to 3.3 inches (5.3 to 8.4 cm) long and 1.5 to 3 inches (3.8 to 7.6 cm) wide; hind tracks 2 to 3.2 inches (5.1 to 8.1 cm) long and 0.5 to 2.4 inches (1.3 to 6 cm) wide

Range: Western mountains of North America from southern Canada to Mexico

Habitat: Mountains

Diet: Alpine plants and flowers

FUN FACT

Bighorn sheep usually live in groups of about 8 to 10 members. Sometimes herds can grow to more than 100.

BLACK BEAR *(URSUS AMERICANUS)*

Black bears dig in the earth for roots and grubs and grasp fish from cold waters with their powerful, sharp claws. These claws show up in their enormous paw tracks. The largest toe on a black bear's foot appears on the outside. Because black bears hibernate in winter, it is unusual to find their tracks in snow. In warmer months, tracks often appear in mud or firm sand along waterways. Generations of bears travel the same trails year after year.

HOW TO SPOT

Track Size: Front tracks 3 to 4 inches (7.6 to 10.2 cm) wide and 4 inches (10.2 cm) long plus a 1.2-inch (3 cm) claw; hind tracks 3.5 to 4 inches (8.9 to 10.2 cm) wide and 7 inches (17.8 cm) long plus a 1.2-inch (3 cm) claw

Range: United States, Canada, and northern Mexico

Habitat: Mountains and forests

Diet: Roots, berries, fish, insects, larvae, grasses, small animals, and livestock

BLACK-FOOTED FERRET

(MUSTELA NIGRIPES)

The black-footed ferret, also known as the American polecat, is a terrestrial animal that hunts at night. Its long, slender body makes it easy to crawl in and out of the holes and burrows of its primary prey, the prairie dog. Black-footed ferret tracks are like those of other small carnivores. They have five toes splayed in a star shape. Claw marks may be visible, especially in soft ground, but they often merge with the imprints from toe pads.

HOW TO SPOT

Track Size: Front tracks about 1.3 inches (3.3 cm) long and 1.3 inches (3.3 cm) wide; hind tracks about 2.1 inches (5.3 cm) long and 1.3 inches (3.3 cm) wide

Range: Central United States

Habitat: Prairies

Diet: Small mammals, birds, snakes, lizards, frogs, fish, and eggs

Tracks of a European polecat, a close relative of the black-footed ferret

COMMON RACCOON

(PROCYON LOTOR)

The common raccoon is well known for its unusual appearance. It has a black "mask" over its eyes and a bushy tail with black rings. These animals are clever and have strong problem-solving skills. Raccoons can untie knots, unscrew jar lids, lift latch locks, and turn doorknobs. They usually walk with a slow, shuffling gait, but they can move quickly when they need to. They are excellent climbers and swimmers. With five slim fingers, raccoon paw prints sometimes look like human hands, although they do not have opposable thumbs.

HOW TO SPOT

Track Size: Front tracks 1 to 2.5 inches (2.5 to 6.4 cm) wide and long; hind tracks up to 4 inches (10.2 cm) long

Range: Southern Canada and most of the United States

Habitat: Woodlands, farmlands, and urban and suburban areas

Diet: Fish, seeds, nuts, berries, insects, eggs, crustaceans, small birds and mammals, garden produce, and pet food

WASH YOUR HANDS!

Scientists once thought that raccoons washed their food in water before eating. Now it is believed that raccoons dip their hands in water to search for food. They rub what they find with their fingers to pick off bits that they don't want before eating it.

COYOTE *(CANIS LATRANS)*

Neither tracks nor scat are needed to announce when a coyote is nearby. Coyotes make their presence known by yipping and yapping at night. These high-pitched howls are sometimes called the "coyote's song." Coyotes howling in the moonlight will usually stay off in the distance, so for those who want an up-close experience with these regal animals, an examination of their tracks will have to do. Coyote prints are oval shaped. Four toes extend from the footpad. All four toes have claws, but sometimes only the claw marks on the front two toes will show in the track.

HOW TO SPOT

Track Size: About 2.5 to 3.5 inches (6.4 to 8.9 cm) long

Range: United States, Canada, and Mexico

Habitat: Prevalent in all habitats but most prevalent in prairies and deserts

Diet: Small mammals, fruits, birds, eggs, vegetables, insects, and fish

DEER MOUSE

(PEROMYSCUS MANICULATUS)

Deer mice are tiny creatures with little, white feet. Their front feet have four toes and their back feet have five. They are so small and lightweight that it's hard to get a complete track unless they are scurrying on firm sand or wet snow. Tracks will appear faintly on sandy beaches or soft river mud. Deer mice make nests in hollow tree trunks and branches to hide from the owls, foxes, and coyotes that prey on them. They can survive outdoors easily, but if it gets very cold, they sometimes invade human homes.

HOW TO SPOT

Track Size: 0.3 inches (0.8 cm) wide and 0.3 inches (0.8 cm) long

Range: Canada, north-central Mexico, and United States except the Southeast

Habitat: All habitats except deserts; most prevalent in woodlands

Diet: Insects, earthworms, snails, fungi, fruits, and carrion

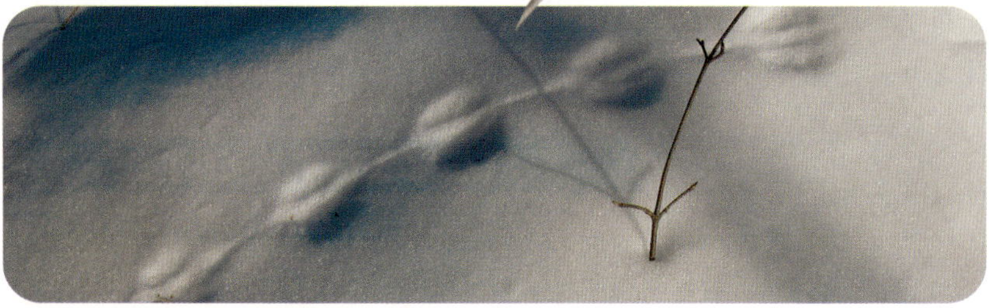

HEALTH HAZARDS

Sometimes animals pose health dangers for humans. Deer mice can transmit a disease to humans called hantavirus. This disease causes serious breathing problems. Special care must be taken when cleaning up the scat of a deer mouse. It is dangerous to touch or breathe in.

DOMESTIC CAT *(FELIS CATUS)*

The domestic cat is a small, furry mammal that is meant to be kept as a household pet. Cats provide companionship for their human caregivers, and they hunt mice and help with pest control. However, not all cats are lucky enough to live in homes. Some domesticated cats live outdoors without human care. These are called feral cats. They are typically the offspring of abandoned or lost pet cats. It is not unusual to find domestic cat tracks in nature. Cat tracks show four toes on the front foot and four toes on the hind foot. Both front and hind paws have claws.

HOW TO SPOT

Track Size: Front tracks 1.5 inches (3.8 cm) long and 1.3 inches (3.3 cm) wide; hind tracks 1.3 inches (3.3 cm) long and 1.3 inches (3.3 cm) wide

Range: North America

Habitat: Pets live in human dwellings and feral cats live outdoors in many types of habitats

Diet: Pet cats eat commercial pet food and can be fed plain, cooked meat. Feral cats hunt small animals and eat from garbage cans

FUN FACT

Domestic cat paw pads help regulate a cat's body temperature and absorb shock when jumping from high places. They also help cats sense vibrations.

DOMESTIC DOG

(CANIS LUPUS FAMILIARIS)

The domestic dog is a descendant of the gray wolf. Through the centuries, these wild dogs were domesticated so that they would be safe companions for humans. Domesticated dogs are often affectionate and can develop close relationships with their human caretakers. Walking on a trail with a dog can be a great way to get exercise and enjoy the outdoors, and it is not unusual to find dog tracks in nature. Tracks often show four toes with claws and a footpad. The pads have a rough texture and are filled with sweat glands, which help dogs grip the ground.

FUN FACT

Some domestic dog breeds, such as Labrador retrievers and Portuguese water dogs, have webbed paws. This helps these dogs swim.

HOW TO SPOT

Track Size: Paw size depends on breed. Large dogs have paws like wild dogs, about 4 inches (10.2 cm) wide

Range: North America

Habitat: Pets live in human dwellings and feral dogs live in several habitats

Diet: Pet dogs eat commercial pet food and can be fed some plain, cooked meat and certain fruits and vegetables. Feral dogs hunt small animals and eat from garbage cans

ERMINE *(MUSTELA ERMINEA)*

With its beautiful white coat, this short-haired weasel blends in perfectly with its snowy surroundings in winter. In the summer, ermines lose their white coat. It is replaced with deep brown fur. Ermines have a musky odor, and people sometimes smell their unpleasant aroma near their dens in warmer months. Ermines are bounders, hopping forward with their front and hind feet together. They leave five-toed prints side by side in a zigzag pattern. Sometimes as they hunt, they'll make graceful leaps after mice, chipmunks, and birds. Then their prints will lose the zigzag pattern.

HOW TO SPOT

Track Size: 0.7 inches (1.8 cm) wide and 0.8 inches (2 cm) long

Range: Northern Canada, Sierra Nevada mountain range, and Alaska in the United States

Habitat: Boreal forests and Arctic tundra

Diet: Small mammals and insects

BIRD BULLIES

Ermine weasels are native to northern North America and Eurasia. They were introduced to New Zealand in the 1880s to help control the rabbit population. Instead, they hunted native birds. As a result, some native bird species in New Zealand are now extinct.

HARBOR SEAL *(PHOCA VITULINA)*

Harbor seals are marine mammals that spend a large part of their time on land. When a harbor seal leaves the water, it is called "hauling out." Seals haul out on average about seven hours per day. Harbor seals have five webbed toes with claws on their front flippers and five toes on their back flippers. Seals use their hind flippers to undulate. When seals undulate, they propel forward in a side-to-side motion. This is how they move on land. Along with its flippers, a seal's belly print is prominent in the tracks it makes in the sandy soil on shore.

FUN FACT

Harbor seals molt, or shed their hair, for about two months each summer. During molting season, seals will haul out for about 12 hours per day.

HOW TO SPOT

Track Size: Flippers including claws are 15 to 16 inches (38.1 to 40.6 cm) wide

Range: North American coastlines and some freshwater lakes in Canada and Alaska in the United States

Habitat: Coastal waters, offshore rocks, and sandbars

Diet: Fish, crabs, lobster, and squid

HORSE *(EQUUS CABALLUS)*

Some horses are wild, but most live on farms or ranches and are cared for by humans. Horse tracks are often found on nature trails where riders take them for recreation and relaxation. Horses are ungulates, so they have hooves. They also have a footpad, called a "frog," that provides protection and absorbs shock as the horse trots or runs. Horses that are cared for by humans sometimes wear horseshoes attached to their hooves. The shoes and the footpads show up in a horse's tracks.

HOW TO SPOT

Track Size: 2.5 to 4 inches (6.4 to 10.2 cm) long

Range: North America

Habitat: Wild horses live in open grasslands and horses kept by humans live in barns, stables, and pastures

Diet: Grasses and grains

FUN FACT

Horseshoes are nailed to the horse's hooves. This does not hurt because the hooves are made of keratin and do not have nerve endings that would cause the horse to feel pain.

JAGUAR *(PANTHERA ONCA)*

The jaguar is the largest cat species native to North America and the third largest wild cat in the world. Jaguars are excellent swimmers and can cross wide, fast-running rivers. Of all the powerful wild cats, the jaguar has the strongest jaw muscles. Its jaw is twice as strong as a tiger's. Jaguars' toes fan out from large footpads. Above the M-shaped pad, the toes make four oval-shaped prints. Claw marks do not show up in tracks because jaguars retract their claws when they walk.

HOW TO SPOT

Track Size: 4 inches (10.2 cm) long and 4.8 inches (12.2 cm) wide

Range: Southern Mexico

Habitat: Near lakes, rivers, and wetlands

Diet: Reptiles, mammals, fish, birds, and caimans

MEADOW VOLE

(MICROTUS PENNSYLVANICUS)

The meadow vole is a short, stocky rodent sometimes called a field mouse. Meadow voles tunnel underground in search of grass stems, roots, and seeds to eat. In winter, when grass is dormant, they eat tree roots and bark. Because they live mostly underground, it can be hard to observe a meadow vole. Ruts of soil show their tunnel patterns. Meadow voles make snow tracks as well. Their tiny paws have strong claws that help them tunnel through hard earth.

HOW TO SPOT

Track Size: Front tracks about 0.3 inches (0.8 cm) wide and 0.3 inches (0.8 cm) long; hind tracks about 0.3 inches (0.8 cm) wide and 0.5 inches (1.3 cm) long

Range: Canada and northern and eastern United States

Habitat: Meadows, lowland fields, grassy marshes, and areas along rivers and lakes

Diet: Grasses, sedges, seeds, tubers, roots, insects, and other invertebrates

MOOSE *(ALCES ALCES)*

Moose are among the largest members of the deer family. Surprisingly, these enormous animals are strong, graceful swimmers. They can swim for about 10 miles (16.1 km) without stopping. They are even more powerful on land. They walk 5 to 10 miles (8 to 16.1 km) each day and can run as fast as 35 miles per hour (56.3 kmh) in a short sprint. Moose have two-toed hooves. Their tracks show two big toes in an upside-down heart shape. Their legs are so tall that their tracks can be more than 6 feet (1.8 m) apart.

HOW TO SPOT

Track Size: 4.5 to 7 inches (11.4 to 17.8 cm) long

Range: Canada and northern United States

Habitat: Forests

Diet: Leaves, trees, bugs, and aquatic vegetation

FUN FACT

Moose are herbivores. They eat only plants and no meat at all. They can eat up to 70 pounds (31.8 kg) of vegetation every day. The common name comes from the Native American Algonquian languages.

MULE DEER *(ODOCOILEUS HEMIONUS)*

The mule deer gets its name from its large, floppy ears, which look like they belong on a mule. Mule deer are hunted by predators, including bobcats, coyotes, pumas, and wolves. When cornered, mule deer have a trick to escape. A jumping gait called "stotting" allows them to leap high in the air with all four legs at once. Mule deer, like other members of the deer family, have cloven hooves, which means they have two toes on each foot. These deer move quickly and nimbly in their rocky habitat.

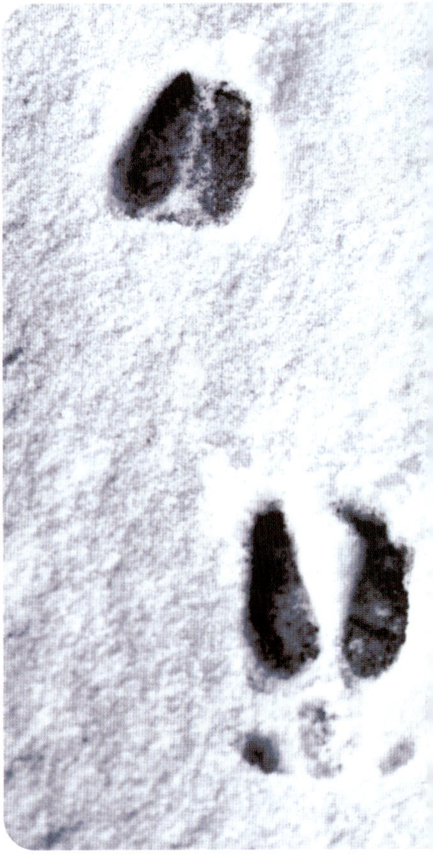

HOW TO SPOT

Track Size: About 2.5 inches (6.4 cm) long

Range: Western United States, western Canada, and Mexico

Habitat: Arid, rocky regions

Diet: Grasses, leaves, twigs, tree bark, and farm crops

MUSK OX *(OVIBOS MOSCHATUS)*

The musk ox is a large, long-haired mammal with a stocky body and a stubby tail. Both male and female musk oxen have horns, but the horns of males, or bulls, are larger and heavier than those of females. A bull's horns span almost its entire forehead. These animals survive and thrive in the extreme cold of the Arctic north. Their coat is long and coarse on the outer layer. And they have a layer of shorter, fine hair underneath to keep them extra warm. Musk oxen are ungulates. They have four two-toed hooves that are all about the same size.

HOW TO SPOT

Track Size: 4.5 inches (11.4 cm) long and 4.5 inches (11.4 cm) wide

Range: Alaska in the United States and northern Canada

Habitat: Arctic tundra

Diet: Arctic grasses, willows, woody plants, lichens, and mosses

BEARDED BOVINES

Musk oxen have long hair. Some of the Inuit peoples in Alaska and northern Canada speak a language called Iñupiaq. In this language, the musk ox is called *umingmak.* This word means "the animal with skin like a beard."

NINE-BANDED ARMADILLO

(DASYPUS NOVEMCINCTUS)

Covered with bony plates, the nine-banded armadillo is an odd-looking creature. With large, pointed ears and a long snout, this animal is closely related to anteaters and sloths. In fact, its tongue is sticky like an anteater's, allowing it to consume large amounts of insects. Armadillos often live near human-populated areas, and it is not unusual to find their tracks in soft sand and mud. Nine-banded armadillos leave four long, clawed toe prints in their tracks. A track of their scaly tail appears between the footprints.

HOW TO SPOT

Track Size: 2.4 to 3.2 inches (6.1 to 8.1 cm) long and 3 to 3.8 inches (7.6 to 9.7 cm) wide

Range: Mexico and southern United States

Habitat: Brush, woods, and grasslands

Diet: Insects, plants, and carrion

NORTH AMERICAN PORCUPINE

(ERETHIZON DORSATUM)

The quill-covered North American porcupine has a waddling gait. Its most famous defense mechanism is its sharp quills. The sharp spikes fall away easily when touched, causing painful injuries to people and pets. Some porcupine tracks show that the animal's front paws have four claws along with a thumb. Their back paws have five claws. Their feet don't always show up in the track, but it is obvious when this animal has recently ambled by. As the quills drag along, the ground looks like it has been swept with a bristly broom.

FUN FACT

Porcupines have about 30,000 quills on their body. The quills grow back after they've fallen out and embedded in the porcupine's victim. The word *porcupine* means "quill pig" in Latin.

HOW TO SPOT

Track Size: Front tracks 1.3 to 2 inches (3.3 to 5.1 cm) long and 1.1 to 1.6 inches (2.8 to 4.1 cm) wide; hind tracks 1.4 to 2.7 inches (3.6 to 6.9 cm) long and 1 to 2 inches (2.5 to 5.1 cm) wide

Range: Canada, northern Mexico, and northeastern and western United States

Habitat: Forests, grasslands, desert shrublands, and tundra

Diet: Tree bark, buds, leaves, shed antlers, and bones of dead animals

NORTH AMERICAN WOLVERINE

(GULO GULO LUSCUS)

Though its name implies that it might be a canine, like a wolf, the North American wolverine is a member of the weasel family. It has a thick body, short legs, short ears, and a broad, flat head. These animals live in extremely cold climates. To help keep them warm in their Arctic surroundings, they have long, dense fur. Like humans and bears, wolverines walk on the soles of their five-toed feet. They have curved claws they can partly retract that allow them to climb trees easily. Their toes spread out as they walk, and their claws and fat footpads appear in their tracks.

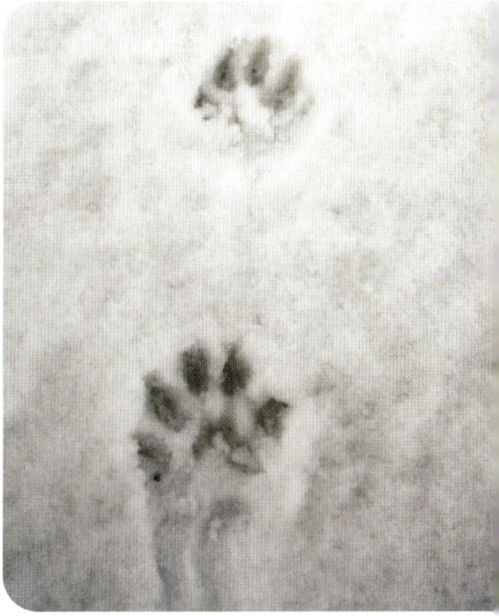

HOW TO SPOT

Track Size: 4.5 to 7.5 inches (11.4 to 19.1 cm) long
Range: Canada, Alaska, and northwestern United States
Habitat: Tundras, boreal forests, and mountains
Diet: Reindeer, small mammals, hares, foxes, birds, frogs, and berries

OCELOT *(LEOPARDUS PARDALIS)*

Compared with other wildcats, the short-legged ocelot is not very big: it's only 3 feet (0.9 m) long. During the day, ocelots' spotted coats give them camouflage in leaves and trees. They are nocturnal and tend to hunt alone at night. Sometimes they stroll along wooded trails searching for prey. Other times they lie in wait, ready to pounce. They have large, padded feet with five toes on their front paws and four toes on their rear paws. Their sharp claws are retractable, so the claws do not always show up in their tracks.

HOW TO SPOT

Track Size: Front tracks 4.2 inches (10.7 cm) long and 3.7 inches (9.4 cm) wide; hind tracks 3.1 inches (7.9 cm) long and 3 inches (7.6 cm) wide

Range: Mexico and southern Texas in the United States

Habitat: Tropical rain forests

Diet: Small mammals, reptiles, birds, and rodents

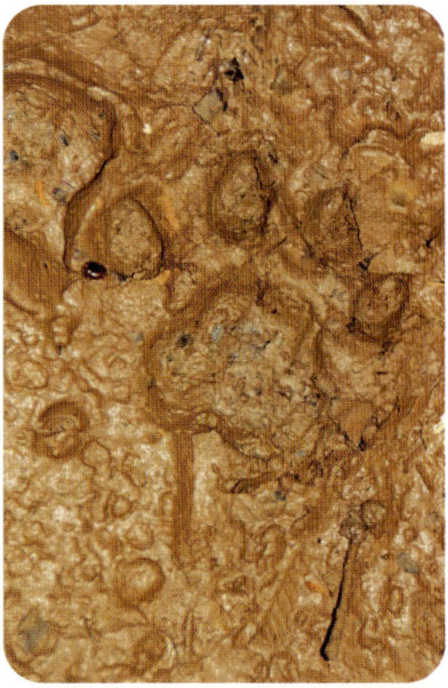

FUN FACT

Ocelots have a rough, sandpapery tongue. This helps them lick every morsel of meat from their prey.

POLAR BEAR *(URSUS MARITIMUS)*

Polar bears have enormous bodies and shaggy, white coats. Though they are fierce hunters, these bears can also be playful and curious. Polar bears have a plodding gait. They walk slowly and rarely break into a run. As patient hunters, they lie in wait at breathing holes for seals to pop up from the water. Polar bears have the largest of all bear tracks. Footpads, toe pads, and claw marks are present in the tracks. Sometimes brush marks made by fur appear as well.

HOW TO SPOT

Track Size: Front tracks 5.5 inches (14 cm) long and 9 inches (22.9 cm) wide; hind tracks 13 inches (33 cm) long and 9 inches (22.9 cm) wide

Range: Canada and Alaska in the United States

Habitat: Sea ice floes on the Arctic Ocean

Diet: Seals

PUMA *(PUMA CONCOLOR)*

Pumas are large wildcats. They are stealthy hunters that prey mostly on deer. Pumas are agile animals. They are strong swimmers, nimble climbers, and fast runners. They can reach speeds of up to 50 miles per hour (80.5 kmh). When these cats run, their claw marks might show in their tracks, which means they are using their claws to help propel them to run faster. If they are simply strolling along slowly, the claw marks do not usually appear in their tracks. Puma paws have four teardrop-shaped toes and three separate lobes in their footpads.

HOW TO SPOT

Track size: Front tracks 2.7 to 4 inches (6.9 to 10.2 cm) long and 2.8 to 4.8 inches (7.1 to 12.2 cm) wide; hind tracks 3 to 4 inches (7.6 to 10.2 cm) long and 2.5 to 4.8 inches (6.4 to 12.2 cm) wide

Range: United States, Mexico, and Canada

Habitat: Mountains, forests, deserts, and wetlands

Diet: Deer, small animals, and insects

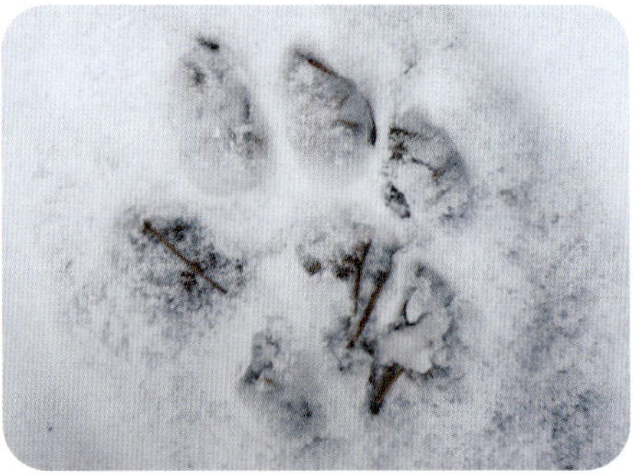

WHAT'S IN A NAME?

Pumas have about 40 different names in the English language, including panther, cougar, red tiger, and mountain lion. Unlike actual lions, pumas cannot roar. They can purr, though! Most scientists use the name puma for this wildcat.

RED FOX *(VULPES FULVA)*

The red fox gets its name from its reddish, rust-colored coat. Its large, pointy ears help it hear tiny mouse scratches up to 109 yards (100 m) away. When foxes walk, their back paws land in their front paw prints, and their gait usually leaves tracks that fall in a straight line. Prints that are far apart reveal that the fox was running, either to catch prey or to outrun a predator. Prints are closer together when the fox goes on a more casual, loping walk. The red fox's paw prints feature four toes, four claws, and a triangular heel pad.

HOW TO SPOT

Track Size: 1.7 to 2.5 inches (4.3 to 6.4 cm) long and 1.5 to 2 inches (3.8 to 5.1 cm) wide

Range: United States and Canada

Habitat: Woodlands, rural and suburban neighborhoods, wetlands, and brushy fields

Diet: Small animals, birds, fruits, nuts, eggs, insects, carrion, and garbage

RED SQUIRREL

(TAMIASCIURUS HUDSONICUS)

Red squirrels are a common sight throughout the United States. They are gray or rust colored with white bellies and thick, bushy red tails. They are very busy in summer, collecting and hiding seeds and nuts. Sometimes they forget to dig them up, so they plant trees by accident. Red squirrels are bounders. When they bound or hop, their smaller front feet land first, followed by their larger hind feet that pass around the front feet to land in front of them. Four front toes and five back toes will show up in their tracks, but their claws don't usually make an imprint.

HOW TO SPOT

Track Size: Front tracks 0.6 inches (1.5 cm) long; hind tracks 1 inch (2.5 cm) long

Range: Canada, Rocky Mountains, northeastern United States, and Alaska

Habitat: Woodlands and forests

Diet: Seeds and cones of evergreen trees, bird eggs, berries, and fruits

RIVER OTTER *(LUTRA CANADENSIS)*

River otters have long bodies, short legs, and slim tails. They are known for their playful nature and silly voices. When they communicate with one another, they giggle, hiss, chuckle, and shriek. Thanks to their streamlined bodies and webbed toes, they are excellent swimmers. Their tracks can be found near rivers and streams where they make their dens. The tracks show the otters' heel pad and claws with fanned-out toes, although their toe webs do not always show up in the print.

HOW TO SPOT

Track Size: About 3.2 inches (8.1 cm) wide and 4 inches (10.2 cm) long

Range: United States and Canada

Habitat: Rivers, streams, lakes, ponds, and marshes

Diet: Fish

SNOWSHOE HARE

(LEPUS AMERICANUS)

The snowshoe hare has a long body, large ears, and huge hind feet. Its brown fur turns white in winter to help it hide in the snow from predators. Its feet are also extra furry in the winter. The stiff hairs act as snowshoes, keeping it from sinking into the deep snow. Even the soles of its feet have fur to protect it from freezing temperatures. Both front feet and hind feet show up in their tracks, with the hind feet on the outside and the front feet between them. Their five toes can be hard to see in their snow tracks, thanks to their extra-furry feet.

HOW TO SPOT

Track Size: Front tracks 1.7 inches (4.3 cm) long; hind tracks 5 inches (12.7 cm) long

Range: Canada and northern United States

Habitat: Boreal forests

Diet: Herbs, twigs, bugs, bark, and their own feces

FUN FACT

The snowshoe hare's large ears help it hear predators' footsteps and wing flaps from miles away. They know when to run and hide.

STAR-NOSED MOLE

(CONDYLURA CRISTATA)

The star-nosed mole has an odd-looking, sensitive nose. This tiny mole lives in wet soil and tunnels through mud, using its star-shaped nose to feel for worms and bugs. This mole is also an excellent swimmer and will dive into shallow water, feeling along the muddy bottom for food with its unique nose. During the winter months, when the ground is covered with snow, the mole may tunnel so close to the surface that it pushes dirt up into the snow cover. When the snow melts, the mounds of dirt over the tunnels suddenly appear as tracks.

HOW TO SPOT

Track Size: Mud holes 1.5 to 2 inches (3.8 to 5.1 cm) in diameter
Range: Eastern United States and eastern Canada
Habitat: Wetlands, marshes, and other areas with moist soil
Diet: Worms, insects, and other invertebrates

WILD BOAR *(SUS SCROFA)*

Native to Europe and Asia, the wild boar was introduced to North America as a game animal. Game animals are animals that humans hunt for sport. Many of the boars that were released into the wild to be hunted survived and thrived instead. Now they appear in 35 of the 50 United States. Wild boars have smaller hind feet than front feet. When they trot, their weight can cause their hooves to spread in the track, making their feet appear wider than they actually are. The dewclaws of the front feet often show in the track, but those on the back feet rarely do.

HOW TO SPOT

Track Size: 2.8 inches (7 cm) wide and 3.9 inches (10 cm) long

Range: Southern and western United States and northern Mexico

Habitat: Savannas, wooded forests, agricultural areas, shrublands, and marshy swamplands

Diet: Berries, roots, and worms

BOAR BADGES

Wild boars leave many traces. They make muddy holes when they dig and snuffle in the dirt in search of roots and worms. They also create wallows, or small depressions in the soil, near marshes and ponds to cool down and ward off biting insects. After wallowing, wild boars rub their bodies on trees, logs, and fence posts. They leave mud and hair behind.

YELLOW-BELLIED MARMOT

(MARMOTA FLAVIVENTRIS)

The yellow-bellied marmot belongs to the squirrel family and is a close cousin of the groundhog. This rodent, also called a rock chuck, has yellow fur on its belly. The rest of its body is covered with brown fur, and it has a white spot between its eyes. Marmots dig elaborate burrows underground and live in colonies of up to 20 members. They emerge from their burrows only in the morning and evening to eat. Their tracks show round foot pads and four toes on their front feet. Their back feet have five toes. Sometimes claw marks appear but not always.

FUN FACT

The yellow-bellied marmot is nicknamed the "whistle pig" for the high-pitched chirp it makes. It makes this noise to warn other group members about potential threats.

HOW TO SPOT

Track Size: 1.5 inches (3.8 cm) long

Range: Western United States and southwestern Canada

Habitat: Steppes, alpine meadows, forest edges, pastures, and gravel-covered fields

Diet: Plants, eggs, and insects

AMERICAN ALLIGATOR

(ALLIGATOR MISSISSIPPIENSIS)

American alligators use their powerful tails and webbed hind feet to move through the water. They can also run very fast, allowing them to catch prey on land. Because they are cold-blooded, they can often be found basking in the sun to stay warm. American alligator tracks are usually found in the muddy shores of the waterways where they live. They have five toes on their front feet and four toes on their hind feet. Their heavy tails leave a deep line between their footprints.

HOW TO SPOT

Track Size: Front tracks about 5 inches (12.7 cm) long and 4 inches (10.2 cm) wide; hind tracks about 8 inches (20.3 cm) long and 5.3 inches (13.5 cm) wide

Range: Southern and eastern United States

Habitat: Freshwater rivers, swamps, marshes, and lakes

Diet: Fish, birds, turtles, snakes, mammals, insects, amphibians, snails, and worms

COMMON SNAPPING TURTLE

(CHELYDRA SERPENTINA)

The common snapping turtle is a large, freshwater turtle that spends so much time in water that it is rare to find tracks. They bury themselves in mud, and sometimes only their eyes and nostrils are exposed. Common snapping turtles can stay underwater for up to three hours. Like most turtles, female common snapping turtles leave their watery habitat to lay their eggs. The best time to look for tracks is in spring and summer when females leave the water to lay their clutches. Their tracks are oval, with five toes showing on the inside of the oval.

HOW TO SPOT

Track Size: 1 inch (2.5 cm) wide

Range: Eastern North America from southern Canada to the Gulf of Mexico

Habitat: Shallow ponds, shallow lakes, and streams

Diet: Invertebrates, fish, frogs, small snakes, small turtles, birds, and small mammals

FUN FACT

Scientists have found that the common snapping turtle has a unique internal compass. Hatchling turtles that have been captured and moved know where the nearest water source is, and they move in that direction once they are released.

GILA MONSTER

(HELODERMA SUSPECTUM)

The Gila monster is the largest lizard native to the United States. These reptiles live most of their lives underground and emerge only to eat or bask in the sun. Gila monsters store fat in their thick tails, so they do not need to eat very often. Scientists believe that they only need to eat three or four large meals per year. Because they spend so much time underground, it is rare to see Gila monster tracks. Gila monsters swing their tail from side to side when they walk. Their wide feet have five clawed toes.

HOW TO SPOT

Track Size: 3 inches (7.6 cm) long

Range: Southwestern United States and northwestern Mexico

Habitat: Rocky foothills

Diet: Mammals, birds, and eggs

FUN FACT

The Native American Apache people once believed that the Gila monster had such terrible breath that it could kill a person. However, this lizard's breath isn't always that bad. If it eats carrion, its breath might smell foul, but if it eats fresh food, its breath is not terrible.

SIDEWINDER *(CROTALUS CERASTES)*

The venomous sidewinder is nicknamed the "horned rattler" for the upturned scales over each eye. Scientists believe that these extra scales protect the snake's eyes from the desert sand. Its actual name, the sidewinder, comes from its unique way of moving. Unlike many snakes that glide forward with their bellies on the ground, the sidewinder lifts its belly off the hot sand as it moves. It tosses its head and tail in S-shaped loops and moves sideways across the sand. Sidewinder snakes can slither up to 18 miles per hour (29 kmh). This movement leaves unique J-shaped tracks.

HOW TO SPOT

Track Size: 17 to 30 inches (43.2 to 76.2 cm) wide

Range: Southern United States and northern Mexico

Habitat: Deserts, open flats, and rocky hillsides

Diet: Rodents and lizards

ATLANTIC PUFFIN

(FRATERCULA ARCTICA)

Atlantic puffins live in colonies on coastal cliffs and islands in the northern Atlantic Ocean. These birds return to the same site year after year to raise baby pufflings. Atlantic puffins are excellent divers. They catch their prey by diving from the surface and using their wings to propel them underwater. Their poop, called guano, reflects their fishy diet. It emerges in a light-colored, mostly liquid stream and has a strong fishy odor. Rocks where puffins roost are often covered with splatters of their guano.

FUN FACT

Puffin guano plays an important role in fertilizing and nourishing plant life in the birds' cold and rocky habitats. Because puffins eat only fish, their guano is rich in nutrients, such as nitrogen, phosphorus, and potassium. Plants in harsh climates need these nutrients to thrive.

HOW TO SPOT

Scat Size: Unknown
Range: Northern Atlantic and Pacific Oceans
Habitat: Rocky cliffs and coastal islands
Diet: Small fish

BARRED OWL *(STRIX VARIA)*

Barred owls are large, nocturnal birds. They have brown and white feathers, dark eyes, and no visible ear tufts. Their broad wingspans can reach up to 4 feet (1.2 m) wide. At night in the dark forests, their distinctive hooting calls are heard. Some people think their hoots sound like the question "Who cooks for you?" Barred owls expel most of their waste in the form of regurgitated pellets. Because of this, the volume of their feces is relatively small. Small dark droppings are released along with splotches of white uric acid.

FUN FACT

Both male and female barred owls bob their heads and wings and call to each other to attract mates. Male barred owls sometimes even feed their mates.

HOW TO SPOT

Scat Size: Unknown

Range: Southern Canada and eastern and northwestern United States

Habitat: Woodlands, wooded areas near rivers, and wooded swamps

Diet: Rats, shrews, voles, birds, reptiles, amphibians, crustaceans, fish, and earthworms

BRANT GOOSE *(BRANTA BERNICLA)*

There are four types of brant geese in North America. In many ways, brant geese are not like other geese. They do not fly in straight lines or in a V-shape when they migrate. Instead, they fly in unorganized groups. Compared with other geese, the brant goose is small. With a black head, dark brown wings, a gray underbelly, and a white ring around its neck, this bird is often mistaken for a duck. When it comes to poop, brant geese are similar to other geese. Their droppings are black to dark green, depending on diet. They feature a dollop of white uric acid at one end.

HOW TO SPOT

Scat Size: About 0.3 inches (0.8 cm) wide

Range: Arctic coast, Pacific coast, southeastern Canada, and coastline of northeastern United States

Habitat: Beaches, bays, estuaries, and Arctic tundra

Diet: Aquatic weeds, grasses, sedges, and mosses

BROAD-BILLED HUMMINGBIRD

(CYNANTHUS LATIROSTRIS)

The broad-billed hummingbird is brightly colored with a dark tail and long red bill. These tiny birds have big appetites and can eat up to half their body weight in bugs and nectar. They feed every 10 to 15 minutes, visiting up to 1,000 to 2,000 flowers throughout the day. Like all birds, hummingbirds urinate and defecate simultaneously through the same opening. It is common to see them pooping and peeing midair. Hummingbird poop is typically sticky and watery due to their frequent consumption of sweet liquid nectar. Thanks to the nectar, their poop smells sweet too!

FUN FACT

Hummingbird chicks place their rump vents over the side of their nest and "shoot" their tiny poops out. This keeps their home clean and sanitary.

HOW TO SPOT

Scat Size: About 0.03 inches (0.1 cm) long
Range: Mexico and southwestern United States
Habitat: Meadows and grasslands
Diet: Nectar, aphids, insects, and spiders

BROWN PELICAN

(PELECANUS OCCIDENTALIS)

Though the brown pelican is a large seabird, it is the smallest species of pelican. These social birds nest in low trees or depressions on the ground. A brown pelican has a long, straight bill with a large throat pouch. When the pelican dives into the sea, the pouch scoops up fish and water. It then drains the water from the pouch to swallow the fish. Pelican poop is like that of other waterbirds. Their strongly scented off-white or light-gray poop is a combination of solid and liquid waste and is usually soft and mushy.

FUN FACT

Brown pelicans are strong, skilled swimmers. Some young pelicans have been observed swimming up to 3 miles per hour (4.8 kmh), even when barely able to fly.

HOW TO SPOT

Scat Size: Runny splotches are about 3 to 5 inches (7.6 to 12.7 cm) wide

Range: Eastern and western coasts of the United States and Mexico

Habitat: Beaches, small islands, and sandbars

Diet: Small fish and invertebrates

CANADA GOOSE *(BRANTA CANADENSIS)*

The Canada goose is the most common goose species in North America. These birds are easily recognizable, thanks to their black heads and necks, white chins, and brown bodies. Canada geese are found near bodies of water such as lakes, rivers, and ponds, where they feed on aquatic plants and insects. They are also known to eat grains and grasses. They sometimes forage on farmland. Goose droppings are tubular in shape and may be in a range of colors, including green, white, or dark brown.

HOW TO SPOT

Scat Size: About 0.3 inches (0.8 cm) wide

Range: United States and Canada

Habitat: Freshwater rivers, lakes, ponds, and streams

Diet: Grasses, sedges, aquatic plants, seeds, berries, and grains

DUSKY GROUSE

(DENDRAGAPUS OBSCURUS)

The dusky grouse is a game bird that can fly, but it builds its nest on the ground under shrubs and low-lying trees. These birds are on the large side compared with other grouse species but still only weigh about 2 to 3 pounds (0.9 to 1.4 kg). Their brown-and-gray mottled feathers help them blend in with their forest surroundings. Dusky grouse scat can be found on the forest floor where they live or near their nesting sites. Droppings are long, curved cylinders, usually black or green.

HOW TO SPOT

Scat Size: 0.3 inches (0.8 cm) in diameter

Range: Western United States and western Canada

Habitat: Forests

Diet: Buds, leaves, and insects

GOLDEN EAGLE *(AQUILA CHRYSAETOS)*

The golden eagle is a powerful bird of prey. Its broad, rounded wings and black-brown to dark brown feathers are striking, but it's the bird's golden head and neck that give it its name. Golden eagles are skilled hunters and often work in pairs. Their sharp talons are used to kill and carry the prey. The beak is used only for eating. After a successful hunt and meal, the golden eagle can go for several days before needing to eat again. They excrete gray, white, or brown watery waste through their cloaca, also known as a vent.

HOW TO SPOT

Scat Size: Up to about 3.7 inches (9.4 cm) long

Range: United States, Canada, and northern Mexico

Habitat: Tundra, grasslands, forests, woodlands, deserts, and canyons

Diet: Marmots, hares, mice, martens, foxes, deer, carrion, and sometimes birds

GIZZARD GUTS

Because they lack teeth, all birds have a gizzard in their digestive system where food is digested. Birds who eat soft food have less-developed gizzards. Carnivorous birds such as the golden eagle need strong gizzards to grind up prey into tiny pieces. Pieces that cannot be digested, such as bone and fur, form pellets. Consisting of fur, bones, or feathers, pellets are "vomited" through the mouth in the morning after the bird has eaten.

MOURNING DOVE

(ZENAIDA MACROURA)

The mourning dove is found across North America. These doves have slender tails and small heads. They can often be seen perched on telephone wires or foraging for seeds on the ground. When taking off, they produce a distinct whistling or whinnying sound with their wings. Mourning doves have soft, drawn-out calls that make them sound sad. The calls are often heard in the early morning or evening. They eat mostly seeds, and their droppings are round pellets that consist of dark, solid waste. Sometimes there is a dollop of white uric acid on top.

HOW TO SPOT

Scat Size: 0.2 inches (0.5 cm) in diameter
Range: United States, Mexico, and southern Canada
Habitat: Forest clearings, farmlands, suburbs, prairies, and deserts
Diet: Seeds

TRUMPETER SWAN

(CYGNUS BUCCINATOR)

The trumpeter swan is the largest waterbird native to North America. Its broad wingspan can reach up to 8 feet (2.4 m). These graceful birds have white feathers and long necks. Their unique call sounds a bit like a trumpet blast. They are found in wetland areas, such as lakes, ponds, and rivers, where they feed on aquatic vegetation and small aquatic animals. Swans are monogamous partners. They form pair bonds that can last for many years. Droppings are soft tubes, ranging in color from brown to green.

HOW TO SPOT

Scat Size: 3 to 4 inches (7.6 to 10.2 cm) long

Range: Northwestern United States and western Canada

Habitat: Lakes, ponds, and rivers

Diet: Aquatic vegetation and small aquatic animals

AMERICAN BADGER *(TAXIDEA TAXUS)*

The short-legged, stout-bodied American badger uses its sharp claws to dig tunnels and burrows. Other burrowing animals, such as squirrels and gophers, are the badgers' main meal, so they spend quite a bit of time hunting underground. They do not defecate in their own dens. Instead they poop in shallow pits, called latrines, outside their burrows. They often cover their waste, so it is hard to find. Their scat is sausage shaped and can be firm or soft. It will be slimy and dark if they've eaten a lot of worms, and it has a sweet, musky smell.

HOW TO SPOT

Scat Size: 1.3 to 1.7 inches (3.3 to 4.3 cm) wide
Range: Western, central, and northeastern United States, northern Mexico, and south-central to southwestern Canada
Habitat: Desert shrublands, grasslands, and tundra
Diet: Small mammals, birds, and insects

AMERICAN PIKA *(OCHOTONA PRINCEPS)*

American pikas may look cute, but they are one of the toughest animals in North America. They are among the few mammals that can survive their entire lives in harsh and windy mountainous terrain. These small, rodent-like creatures have short, stout bodies with big, round ears and no visible tail. They are herbivores, and their scat is found in a pile of small, black pellets. Often, the scat will be covered with white uric acid.

HOW TO SPOT

Scat Size: 0.07 to 0.1 inches (0.2 to 0.3 cm) in diameter
Range: Western United States and western Canada
Habitat: Mountains and rock fields
Diet: Grasses, forbs, and small shrubs

COLOR CHANGES

American pikas' brown-and-black coats help them blend in with the rocks in their alpine habitat. And their thick fur keeps them warm in the winter. During the summer, they shed their thick coat for a lighter one, but it is still thick enough to protect them from the elements. Pikas can overheat in warm temperatures.

AMERICAN PINE MARTEN

(MARTES AMERICANA)

As their name suggests, American pine martens live in pine forests, where they make their dens in tree hollows and abandoned burrows. They look similar to their smaller weasel cousins. Pine martens are excellent climbers, can travel under snow for long distances, and are strong swimmers. When defecating, martens wriggle their hips, resulting in twisted, coiled poop that looks like fox scat but smaller and with a sweeter smell. Remnants of animal parts, berries, and leaves are often visible in their scat.

HOW TO SPOT

Scat Size: 2 to 4 inches (5.1 to 10.2 cm) in length; 0.3 to 0.6 inches (0.8 to 1.5 cm) in diameter

Range: Canada and northeastern and western United States

Habitat: Coniferous forests

Diet: Small mammals, eggs, insects, fruits, leaves, and grasses

ANGUS COW *(BOS TAURUS)*

Angus cattle are domesticated, meaning they are raised and cared for on farms and ranches. Though the breed originated in Scotland, these cows are found in abundance in North America. Most are solid black, but some have patches of white on their underbellies or udders. Cow droppings, or dung, from Angus cows are round or oval and have a moist, firm texture. Fresh dung may be soft and mushy, but it gradually dries out and hardens over time. Cow dung is an important resource on farms and ranches. It is commonly collected and used as a fertilizer for crops and pastures.

HOW TO SPOT

Scat Size: 6 to 8 inches (15.2 to 20.3 cm) in diameter

Range: United States and Canada, including the major beef-producing regions of Texas, Nebraska, Kansas, Alberta, and Saskatchewan

Habitat: Pastures and open ranges

Diet: Hay, grasses, and legumes

BIG BROWN BAT *(EPTESICUS FUSCUS)*

The big brown bat gets its name from the color of its fur. It ranges in color from reddish brown to dark brown. These bats are social and live and travel in colonies. As flying mammals, bats rarely leave tracks. Their leg bones and muscles are part of their wings, making it easier for them to fly than to walk. Their feet can turn inward, which allows them to grasp objects such as twigs and branches. Bat poop, also known as guano, is similar in size to a grain of rice and is odorless. Unlike rodent droppings, bat droppings don't harden and may include undigested insects.

HOW TO SPOT

Scat Size: 0.2 to 0.3 inches (0.5 to 0.8 cm) long

Range: United States, Canada, and Mexico

Habitat: Forests, caves, rock crevices, old buildings, bridges, mines, and trees

Diet: Insects

FUN FACT

Big brown bats often live in forests where they can hide in trees. They are also often found in suburban neighborhoods, living in attics, barns, and other buildings.

BOBCAT *(LYNX RUFUS)*

Bobcats have short, stout bodies and long legs with large paws. The fur on their ears sticks out in tufts. Bobcats are solitary animals and mark their territory with their scat and urine. If a bobcat comes across a freshly laid pile of excrement or a tree that has been scratched and squirted with urine, they know to look for another hunting ground. A bobcat's tube-shaped scat is black or brown. Sometimes bobcat excrement appears near where people live and is mistaken for dog droppings. The difference is that bobcat scat usually contains the fur or bones of their prey.

HOW TO SPOT

Scat Size: About 3 to 5 inches (7.6 to 12.7 cm) long

Range: United States, southern Canada, and northern Mexico

Habitat: Forests, swamps, deserts

Diet: Rabbits, hares, rodents, deer, reptiles, birds, and domestic animals

BROWN-THROATED SLOTH

(BRADYPUS VARIEGATUS)

The brown-throated sloth's long, curved claws allow it to hang from tree branches, and its limbs are designed for gripping and hanging. Due to these animals' slow metabolism, they spend most of their time still, conserving energy. Brown-throated sloths can hold up to one-third of their body weight in feces. Once per week they make a slow trip down to the base of the same tree and drop dark, round pellets in a hollow made with their tails. Scientists are not sure why they won't poop in their trees, especially because sloths are very vulnerable to predators while on the ground.

HOW TO SPOT

Scat Size: 1 to 3 inches (2.5 to 7.6 cm) in diameter

Range: Southern Mexico and Costa Rica

Habitat: Tropical forests

Diet: Leaves, flowers, and fruits

BROWN RAT *(RATTUS NORVEGICUS)*

Brown rats have a reputation of being dirty, but these rodents are actually quite clean. They groom themselves the way cats do, by licking their fur. Brown rats are social and live in large, family-based colonies. Scientists report that they get lonely and sad when they are alone. Despite their many positive traits, humans are still distressed when they find that rats have invaded their home. Droppings are usually the first sign that rats have moved in. Brown rat scat is black, shiny, and shaped like rice.

HOW TO SPOT

Scat Size: 0.5 to 0.7 inches (1.3 to 1.8 cm) long
Range: Throughout North America
Habitat: Forests and human dwellings
Diet: Fruits, seeds, insects, small mammals, and human garbage

FUN FACT

Brown rats like being tickled. Scientists report that they laugh a high-pitched noise and their ears turn pink when they are tickled in just the right spot.

CALIFORNIA SEA LION

(ZALOPHUS CALIFORNIANUS)

California sea lions are marine mammals. They are adept at leaping out of the water and can often be seen sitting on top of rocks, floating docks, and beams under fishing piers. These animals form colonies of several thousand members. They are noisy animals—their barking is loud!—and they love to play. California sea lions defecate near their communal home, which is called a rookery. The smell of their excrement is often compared to ammonia and can be quite pungent. The texture of their stool can vary from soft and round to formless blobs.

HOW TO SPOT

Scat Size: 2 to 2.5 inches (5.1 to 6.4 cm) in diameter

Range: Southeastern Alaska to the Pacific coast of central Mexico

Habitat: Coastal waters and rocky shorelines

Diet: Fish, crabs, and squid

CANADA LYNX *(LYNX CANADENSIS)*

The Canada lynx is found across the northern regions of North America. In winter when other animals are hibernating, the snowshoe hare makes up 75 percent of the Canada lynx's diet. The lynx's digestive system efficiently processes meat. Lynx scat is black, tubular, and tapered, containing less undigested material than that of bears or canines. The odor of lynx scat is strong and unpleasant. Like house cats, they tend to cover their scat with dirt or snow, likely to conceal their presence from other animals. They also tend to defecate repeatedly in the same latrine.

HOW TO SPOT

Scat Size: Up to 4 inches (10.2 cm) long and 0.7 inches (1.8 cm) in diameter

Range: Northern United States, Alaska, and Canada

Habitat: Boreal forests

Diet: Snowshoe hares and other small animals

CAPYBARA

(HYDROCHOERUS HYDROCHAERIS)

Capybaras are not native to North America. However, there is a population of wild capybaras in Florida in the United States. These animals, the largest rodents in the world, may have escaped captivity or been intentionally released into the wild. Capybaras are well suited to water, with webbed feet and water-resistant fur. While they sometimes poop on land, they also enter the water to defecate. Fresh capybara scat is dark in color and moist and soft in texture. Because it is rich in protein, Capybaras sometimes eat their own poop!

HOW TO SPOT

Scat Size: About 1 inch (2.5 cm) long

Range: North-central Florida in the United States

Habitat: Wetlands, marshes, and mangrove swamps

Diet: Grasses and aquatic plants

FUN FACT

Capybaras make a variety of vocalizations, including purring, whistling, barking, and clicking sounds. They use these sounds to communicate within their groups and to alert one another to possible danger.

CARIBOU *(RANGIFER TARANDUS)*

Caribou, also known as reindeer, are well known for their unique antlers. Both males and females have these antlers. Males lose theirs in late fall, but they grow back larger each year. Caribou have thick fur coats that keep them warm in their extremely cold surroundings. Splayed hooves provide traction for walking on snow and ice. These animals are herbivorous and eat lichens, mosses, and shrubs. During the winter, their scat has a high fiber content that appears in small, concave pellets. In the summer, it forms as soft clumps due to a diet of more moist plants.

HOW TO SPOT

Scat Size: 0.5 to 0.6 inches (1.3 to 1.5 cm) in diameter

Range: Northern and western United States and central and western Canada

Habitat: Tundra, woodlands, mountains, and forests

Diet: Mosses, grasses, shrubs, leaves, mushrooms, and sedges

DALL SHEEP *(OVIS DALLI)*

Dall sheep have white coats and large horns. These ungulates live in the mountains and have sharp hooves that help them leap and run on the steep and rocky terrain. These sheep are herbivores. They eat a variety of grasses and shrubs during the summer. In the winter, vegetation is harder to find. They eat lichen and other small plants that grow on rocks. Dall sheep scat reflects their herbivorous diet. Sheep droppings are firm and break apart easily. They can be light or dark brown.

FUN FACT

Dall sheep are named for William Healey Dall, an American naturalist who first described the species in the late 19th century.

HOW TO SPOT

Scat Size: 0.5 to 1 inch (1.3 to 2.5 cm) in diameter
Range: Western Canada and Alaska in the United States
Habitat: Ridges, meadows, and steep slopes
Diet: Grasses, sedges, forbs, lichens, and mosses

EASTERN CIVET CAT

(SPILOGALE PUTORIUS)

The eastern civet cat has a misleading name. This animal, also known as the eastern spotted skunk, is closely related to the striped skunk. Unfortunately, the eastern civet cat is far less common than its striped cousin. Farmland replaced the eastern civet cat's wooded habitat, and pesticides killed many of the insects that it eats. The eastern civet cat is an endangered species. In areas where it still exists, this animal helps with pest control, eating mice, rats, and moles that destroy crops. Its scat is like the striped skunk's poop. It is generally dark in color and tube shaped.

HOW TO SPOT

Scat Size: 1 to 2 inches (2.5 to 5.1 cm) long

Range: Western United States

Habitat: Open prairies, brushy areas, and cultivated land

Diet: Bees and wasps as well as their hives, larvae, honey, mice, rats, moles, shrews, ground squirrels, young rabbits, and chipmunks

EASTERN COTTONTAIL RABBIT

(SYLVILAGUS FLORIDANUS)

The Eastern cottontail rabbit can twitch its tiny nose up to 120 times per minute trying to catch the odor of a potential predator. If it senses danger, it will hop and run away. These rabbits can leap distances of 10 to 15 feet (3 to 4.6 m). They can also stand up tall on their hind legs to watch for coyotes, foxes, and other predators. When they are being chased, they leap from side to side so that their scent is hard for their predator to follow. As herbivores, their scat often contains bits of grass and other plant materials. They defecate in piles of rounded balls.

HOW TO SPOT

Scat Size: 0.1 to 0.4 inches (0.3 to 1 cm) in diameter

Range: Southern Canada, central and eastern United States, eastern Mexico, and Central America

Habitat: Fields, woodlands, swamps, and thickets

Diet: Grasses, clover, fruits, vegetables, twigs, brambles, and tree bark

ELK *(CERVUS CANADENSIS)*

Elk are the second-largest members of the deer family in North America, after moose. Elk live in cold climates and have heavy coats of fur. This helps them stay warm in temperatures that can drop to –40° Fahrenheit (–40°C). Elk live in remote habitats that are not often visited by humans, so they are rarely seen. But this animal's tracks and scat can be found in abundance on mountain trails, letting scientists know that their populations are still at healthy levels. Like most types of deer, elk defecate piles of oval-shaped pellets.

HOW TO SPOT

Scat Size: 1 to 1.5 inches (2.5 to 3.8 cm) wide

Range: Southwestern Canada and northwestern United States

Habitat: Upland forests, mountain prairies, tundra, and brushlands

Diet: Grasses, clover, twigs, lichens, bird eggs, bird bones, and acorns

NAME GAME

Around the world, the animal called elk in the United States is known as *wapiti*, a word from one of the Native American Algonquian languages that means "white rump." In Europe, the animal known as moose in the United States is called elk. Moose and elk are two different species of the deer family.

FISHER *(MARTES PENNANTI)*

The fisher is often called "fisher cat," but it is not a cat, nor does it catch fish. This large weasel is a ferocious hunter. Fishers eat many of the animals that share their forest habitat, even prickly porcupines! Fishers know how to avoid the sharp quills. Fisher scat is usually dark and twisted. Fishers can control the amount of feces that they release, and they sometimes release small amounts of scat to mark territory. Fisher scat is often found on raised areas such as tree stumps or at the base of old trees.

HOW TO SPOT

Scat Size: 0.2 to 0.7 inches (0.5 to 1.8 cm) in diameter
Range: Northern United States and southern Canada
Habitat: Forests with thick tree canopies
Diet: Rabbits, snowshoe hares, squirrels, raccoons, mice, reptiles, amphibians, insects, carrion, and occasionally outdoor cats

FUN FACT

Fishers eat mostly animals, but they sometimes eat berries and fruits if they are available. And although they don't hunt for live fish, they will eat dead fish if they find some near a lake or pond.

GRAY FOX *(UROCYON CINEREOARGENTEUS)*

The gray fox is small with grayish-brown fur, a black-tipped tail, and a slender body. During breeding season, they live in bonded family groups and are playful and affectionate with one another. Otherwise, gray foxes tend to be alone. They are nocturnal animals and are active during the night. They eat a variety of foods, including fruits, nuts, insects, and small mammals. Fox scat looks more like cat scat than dog poop. It is dark brown to black, dry, and stringy, and it has a pointy end.

HOW TO SPOT

Scat Size: About 2.5 inches (6.4 cm) long

Range: Southern Canada, United States excluding the northwest, and Mexico

Habitat: Rocky, wooded brush

Diet: Rabbits, rodents, birds, frogs, earthworms, berries, and fruits

GRAY SQUIRREL

(SCIURUS CAROLINENSIS)

Gray squirrels have gray fur and fluffy tails. They also have an excellent sense of smell, which they use to locate food that they've hidden away. These squirrels communicate with one another with sounds and body movements, such as tail flicking. When predators are nearby, gray squirrels will sound warning calls. They eat nuts, seeds, and other vegetation and defecate in small, dark brown, barrel-shaped pellets. Squirrel scat can often be found under trees, near bird feeders, and inside attics and crawl spaces in homes or businesses.

HOW TO SPOT

Scat Size: About 0.3 inches (0.8 cm) long and 0.1 inch (0.3 cm) in diameter

Range: Southeastern Canada and eastern and central United States

Habitat: Forests, woodlands, and urban and suburban areas

Diet: Nuts, seeds, buds, and flowers of trees

FUN FACT

Gray squirrels play an important role in seed dispersal. As winter approaches, squirrels carry their food and bury it in several locations, but they sometimes forget exactly where. Those buried seeds will sprout and grow.

GRAY WOLF *(CANIS LUPUS)*

The gray wolf, also known as the timber wolf, has gray fur, a bushy tail, and a strong, powerful build. Gray wolves are social animals. They live in packs that include parents and their offspring. These wolves are affectionate with one another, and they communicate using a variety of howls, growls, and barks. They work together as they hunt and prey on elk, deer, moose, and bison. Their scat reveals their meals. Wolf scat is often twisted like a cord and may contain animal hair and bone fragments. Sometimes wolf scat is runny after they have eaten fresh meat.

HOW TO SPOT

Scat Size: 0.5 to 1.5 inches (1.3 to 3.8 cm) in diameter

Range: United States and Canada

Habitat: Forests, mountains, tundra, taiga, grasslands, and deserts

Diet: Large and small mammals

GRIZZLY BEAR

(URSUS ARCTOS HORRIBILIS)

The grizzly bear, also known as the North American brown bear, is a large, powerful mammal. It has brown fur, humped shoulders, and a round face. These bears eat a variety of foods, from berries and insects to fish and mammals as large as elk. Their food choices appear in their scat. In spring and summer, grizzly bears eat grasses and insects, and their scat is green and tubular. In late summer and fall, they eat ripe fruits and berries. Their scat becomes looser and appears in plops, often with bits of undigested food visible.

HOW TO SPOT

Scat Size: 5 to 12 inches (12.7 to 30.5 cm) long and 1.5 to 2.5 inches (3.8 to 6.4 cm) in diameter

Range: Northwestern United States and western Canada

Habitat: Mountain forests, subalpine meadows, Arctic tundra, wetlands, grasslands, mixed-conifer forests, and coastal areas

Diet: Berries, roots, grasses, insects, plant bulbs, rodents, elk, moose, mountain goats, and sheep

FUN FACT

Grizzly bears move fast in water and on land. They are strong swimmers and can run at speeds of up to 35 miles per hour (56.3 kmh).

GUINEA PIG *(CAVIA PORCELLUS)*

Guinea pigs are small, domesticated rodents native to South America. They no longer exist in the wild but are widely kept as pets throughout North America. At only about 8 to 10 inches (20.3 to 25.4 cm) long, they are loved for their bright-eyed faces and gentle nature. About the size of a pea and oval-shaped, guinea pig poop is shiny and brown. It is usually firm but not overly hard or dry. Guinea pigs have a second kind of nutrient-rich stool, called cecotropes. Guinea pigs pass them at night and eat them right away.

HOW TO SPOT

Scat Size: 0.2 to 0.4 inches (0.5 to 1 cm) in length
Range: Abundant as pets in North American households
Habitat: Large cages or pens with a solid bottom
Diet: Commercial guinea pig food, fresh hay, leafy greens, vegetables, and fruits

POPULAR PETS

About 1.5 million families keep guinea pigs as pets in the United States. Since most families keep more than one as pets, it is estimated that there are about 3.8 million guinea pigs in the country.

HOARY BAT *(LASIURUS CINEREUS)*

The hoary bat migrates in winter. Rather than spending the cold months hibernating like most bats, it flies to warmer habitats. During the day, these nocturnal mammals hide among the trees in the forests where they live. They hang from one foot and wrap their wings around them, and they look like dry leaves hanging from a tree branch. At night, they hunt moths, mosquitoes, and occasionally other bats. Hoary bat droppings look like mouse droppings. They are small and hard, like black grains of rice.

HOW TO SPOT

Scat Size: 0.1 to 0.5 inches (0.3 to 1.3 cm) long

Range: Throughout North America from northern Canada to southern Mexico

Habitat: Deserts, grasslands, shrublands, and forests

Diet: Moths, beetles, and mosquitoes

FUN FACT

Hoary bats are incredibly fast and agile. They can fly as fast as 13 miles per hour (20.9 kmh) and as high as 8,000 feet (2,438 m) during their long-distance migrations.

HOUSE MOUSE *(MUS MUSCULUS)*

Despite its name, the house mouse does not always live indoors. Some live in nature, where they dig burrows, build nests, or make dens in rocky crevices. A home that has house mice will show many signs. There will be scratching on the walls and under the floors. Usually this happens at night because house mice are nocturnal. They leave gnaw marks on wooden moldings and furniture. And they leave black or dark brown droppings. House mouse scat is rod shaped and pointed on both ends. It can be hard or soft and dry or moist, depending on what the mouse ate.

HOW TO SPOT

Scat Size: Droppings measure about 0.1 to 0.2 inches (0.3 to 0.5 cm) long

Range: Mexico, United States, and most of Canada

Habitat: Human buildings, grasslands, fields, deserts, forests, and savannas

Diet: Insects, seeds, and grains

JAVELINA *(TAYASSU TAJACU)*

The javelina, also known as the collared peccary, is like a pig, but there are differences. Like a pig, it has small ears and a short tail. It also has canine teeth, or tusks. These tusks are like a pig's, except a javelina's teeth are straight, while pig teeth are curved. Javelinas are omnivores and eat a lot. They eat mostly vegetation, but they will also eat small animals and dead birds on occasion. Their scat is the shape of a small pancake if the animal's diet is moist. If the scat is dark green, that means it is fresh, and a javelina is likely nearby.

HOW TO SPOT

Scat Size: 2 to 3 inches (5.1 to 7.6 cm) long
Range: Mexico and southwestern United States
Habitat: Deserts, forests, woodlands, and grasslands
Diet: Fruits, roots, tubers, lizards, dead birds, and rodents

FUN FACT

Javelinas have scent glands on their rumps. They rub their scent on tree stumps to mark their territory.

KODIAK BEAR

(URSUS ARCTOS MIDDENDORFFI)

The Kodiak bear is a large type of brown bear that lives on an archipelago off the coast of Alaska. No other bears live on the Kodiak islands, not even polar bears. They have a varied diet. They eat grasses, leafy plants, and berries. Kodiak bears also eat fish, especially salmon, and they will sometimes eat carrion. Their scat reveals their hearty appetites. Their poop is tube shaped and green or brown when they are eating mostly spring vegetation. It becomes darker and a bit smellier as the bear adds meat to its diet.

HOW TO SPOT

Scat Size: 2 inches (5.1 cm) or more in diameter

Range: Alaska's Kodiak Archipelago in the United States

Habitat: Spruce forests

Diet: Salmon and other fish, grasses, plants, berries, deer, and elk

MOUNTAIN GOAT

(OREAMNOS AMERICANUS)

Despite their stocky bodies, mountain goats are agile climbers. Thanks to their strong legs and broad hooves, they move easily and swiftly on ice and on snow-covered cliffs. This helps them outrun wolves and bears. Because these animals live in remote, hard-to-reach areas, they are difficult to observe. Scientists rely on the goats' scat to learn about them. The goats' small, oval-shaped pellets reveal what they eat and how their digestive system works. There is not much moisture in the mountain goat's diet, and their droppings are often dry and crumbly.

HOW TO SPOT

Scat Size: 0.8 inches (2 cm) long and 0.3 inches (0.8 cm) wide

Range: Western Canada, northwestern United States, and southeastern Alaska

Habitat: Steep mountain ranges

Diet: Grasses, herbs, foliage, twigs, lichens, alpine firs, and other conifers

FUN FACT

Mountain goats, also called rock goats, were not described by scientists until 1816. They are among the least studied animals because they are so difficult to observe.

NORTH AMERICAN WATER VOLE

(MICROTUS RICHARDSONI)

Though they are rarely observed because they spend so much time underground, North American water voles are easy to identify. They have unusually large hind feet that help them move easily through the burrows they create underground near bodies of water. Huge incisor teeth help them dig tunnels and chew through tough roots. When they infest human dwellings, their droppings look like mouse or rat droppings. Their scat looks like rice and can be either green, brown, or reddish in color.

HOW TO SPOT

Scat Size: 0.2 inches (0.5 cm) long

Range: Western United States and western Canada

Habitat: Near streams and lakes in mountainous areas

Diet: Earthworms, underground larvae, insects, and plants

NORTHERN FLYING SQUIRREL

(GLAUCOMYS SABRINUS)

Northern flying squirrels weigh only about 3 to 6 ounces (85 to 170.1 g). Despite their name, these creatures do not have wings. Instead, they have a membrane of skin that stretches from their wrists to their ankles. When they spread their limbs, this membrane allows them to glide up to 150 feet (45.7 m) and even turn in the air. They defecate in a communal latrine, often a tree branch or tree cavity. This behavior helps them avoid predators by minimizing the scent and visibility of their scat. Their scat is cylindrical and elongated, firm, and compact.

HOW TO SPOT

Scat Size: 0.2 to 0.5 inches (0.5 to 1.3 cm)

Range: Canada and northern United States

Habitat: Coniferous and mixed forests

Diet: Seeds, nuts, fruits, and insects

PLAINS BISON *(BISON BISON BISON)*

Plains bison, commonly called buffalo, have horned heads and wide shoulders. While huge, weighing up to 2,400 pounds (1,088.6 kg), they are the smaller of the two bison subspecies in North America. Despite their size, they can jump 6 feet (1.8 m) in the air and run up to 35 miles per hour (56.3 kmh). Plains bison forage for nine to eleven hours per day to fuel their large bodies. Plains bison scat is released as a dark liquid substance that dries into a patty, sometimes called a "buffalo chip." It contains remnants of the plants that make up the animal's diet.

HOW TO SPOT

Scat Size: 10 to 12 quarts (9.5 to 11.4 L) of excrement dropped in many loose, indistinct piles each day

Range: Western United States

Habitat: Open plains, savannas, and grasslands

Diet: Grasses and other vegetation

SEA OTTER *(ENHYDRA LUTRIS)*

Sea otters have everything they need to survive in the frigid ocean waters of their habitat. Their thick, dense fur keeps their body warm. Their strong teeth—twice as strong as human teeth—break through the shells of the clams and mussels they eat. And they eat a lot! Sea otters must eat at least 25 percent of their body weight every day to survive. Animals that eat a lot poop a lot. Sea otter scat is called spraint. Spraint is tube shaped and light in color. It has a jiggly, jelly-like consistency with bits of bone and shell. Sea otters poop both in the ocean and on land.

HOW TO SPOT

Scat Size: 1.5 inches (3.8 cm) in diameter

Range: Coastal waters of California and Alaska in the United States and coastal waters of western Canada

Habitat: Rocky shores, barrier reefs, and dense kelp forests

Diet: Clams, crabs, fish, urchins, mussels, and squid

COOL TOOLS

Sea otters use tools. They hold objects, most often rocks, in their paws and smash open the tough shells of their prey. They even have a handy-dandy place to keep their favorite rock. A small pouch of loose skin under their arm is a perfect spot to store food for later or to hold a helpful tool.

STRIPED SKUNK *(MEPHITIS MEPHITIS)*

The striped skunk is a black-and-white-striped mammal known for its pungent odor. About the size of a house cat, the striped skunk has short legs and sharp claws. Skunks' noses are pointed, and their eyes are small and black. They have scent glands that produce a strong-smelling spray. This spray protects them from predators. Because they are omnivores, their scat reflects their varied diet. Sometimes undigested seeds, berries, and other food particles are found in their dark, tube-shaped droppings.

HOW TO SPOT

Scat Size: 1 to 2 inches (2.5 to 5.1 cm) long and 0.2 to 0.5 inches (0.5 to 1.3 cm) in diameter

Range: North America from southern Canada to northern Mexico

Habitat: Wooded areas, deserts, plains, and urban and suburban environments

Diet: Small animals, fish, insects, spiders, bird hatchlings, eggs, fish, corn, seeds, berries, and fruits

FUN FACT

Striped skunks are the largest species of skunk. Despite their chunky bodies, they can run fast, sometimes up to 10 miles per hour (16 kmh).

TOWNSEND'S MOLE

(SCAPANUS TOWNSENDII)

Townsend's moles are burrowing mammals with poor vision but an advanced sense of touch. Using their sensitive snouts and long, clawed paws, moles dig tunnels underground. One of the most common signs that a Townsend's mole is nearby is the presence of ridges from their tunnels. These tunnels can also cause the ground to feel spongy and loose. Moles also leave behind droppings in the form of small brown pellets. These can often be found in heaps close to the surface of the dug holes.

HOW TO SPOT

Scat Size: 0.2 inches (0.5 cm) long

Range: Northwestern United States and southwestern Canada

Habitat: Open lowlands and wooded areas with moist soil

Diet: Earthworms, underground larvae, and insects

Mole hills

VIRGINIA OPOSSUM

(DIDELPHIS VIRGINIANA)

Opossums have pointed snouts and nearly hairless tails that they use to grasp objects and hang from trees. These animals help with pest control by eating ticks, cockroaches, rats, and mice. Opossums even eat venomous snakes that are dangerous to humans. Opossums tend to use the same area as a latrine repeatedly, so it is common to find a large amount of droppings and urine in one spot. Droppings are smooth on the sides and brown in color. They curl and taper at the ends.

HOW TO SPOT

Scat Size: 1 to 2 inches (2.5 to 5.1 cm) long and about 0.8 inches (1.9 cm) in diameter

Range: Eastern and western United States and southeastern Canada

Habitat: Forests, woods, marshes, and swamps

Diet: Insects, rodents, birds, and carrion

MARSUPIAL MOMS

Opossums are the only marsupials native to North America. Marsupials are animals that give birth to live young that are not fully developed. When an opossum mother gives birth, she keeps her babies in a pouch on her belly until they are about two months old. After they climb out of her pouch, her babies stay with her. They ride on her back until they are about four months old, when they can walk and eat on their own.

WHITE-TAILED DEER

(ODOCOILEUS VIRGIANUS)

As their name suggests, white-tailed deer have fluffy, white tails. These animals startle easily, and people often only catch a glimpse of a white fluff as the deer dashes away. These deer are herbivores. They eat vegetation that is in season, such as leaves and buds in spring, apples and berries in summer, and nuts and acorns in autumn. They defecate in piles of oval-shaped pellets. Fresh pellets are dark green or black.

HOW TO SPOT

Scat Size: 0.2 to 0.5 inches (0.5 to 1.3 cm) in diameter and 0.5 to 1 inch (1.3 to 2.5 cm) long

Range: United States, southern Canada, and Mexico

Habitat: Forests, rain forests, shrublands, and swamps

Diet: Plants, fruits, berries, nuts, and acorns

WOOD BISON *(BISON BISON ATHABASCAE)*

Wood bison are the largest land mammals in North America. Both males and females have curved horns growing outward and upward from the sides of their heads. Their shaggy coats protect them from cold temperatures. They use their large heads and broad shoulder humps to sweep aside snow in the winter to find food. Since they are herbivores, their feces contain undigested bits of vegetation. Wood bison use their feces, which emerges as a liquid and dries into patties, to communicate their presence to other bison and mark their territories.

HOW TO SPOT

Scat Size: 10 to 12 quarts (9.5 to 11.4 L) of excrement dropped in many loose, indistinct piles each day

Range: Northern Canada and Alaska in the United States

Habitat: Boreal forests, woodlands, and tundra

Diet: Grasses, sedges, and other vegetation

COMMON COLLARED LIZARD

(CROTAPHYTUS COLLARIS)

Common collared lizards can run swiftly on their hind legs. Their stride can be up to three times the length of their body. They use their tails and forelimbs to maintain balance as they sprint. Common collared lizards have large heads with strong jaw muscles that allow them to get a powerful grip on their prey. They are known to eat a variety of insects and small animals, including other lizards. To avoid predators, they quickly dive into rock crevices. These lizards expel feces and urine at the same time. Scat is dark with a white cap of uric acid.

HOW TO SPOT

Scat Size: 1.3 inches (3.3 cm) long

Range: South-central United States and northern Mexico

Habitat: Desert grasslands

Diet: Grasshoppers, crickets, and other lizards

DESERT TORTOISE

(GOPHERUS AGASSIZII)

The desert tortoise has very good eyesight. When food is hard to find in its desert environment, its keen eyesight helps it find the vegetation it needs to survive. These tortoises have three kinds of digestive waste: feces, urine, and a substance called urates. Their feces are often a greenish-brown color. Urates is a whitish, creamy substance made up mostly of proteins. Sometimes urates and liquid urine are passed together. For a second chance to digest important nutrients, desert tortoises sometimes eat their own poop.

HOW TO SPOT

Scat Size: 1 to 4 inches (2.5 to 10.2 cm)

Range: Southwestern United States and northern Mexico

Habitat: Sandy flats and rocky foothills

Diet: Grasses, herbs, and cactus fruits and flowers

FUN FACT

Desert tortoises have a special way to keep their bodies hydrated. They can recycle water from their bladder when there is no water for them to drink.

TEXAS BANDED GECKO

(COLEONYX BREVIS)

Texas banded geckos are tricksters. When they are threatened, they act like scorpions by holding their tails above their heads. They have spurs on their tails, and they can break off their tails if they're in danger. They also make a high-pitched squeak when they are disturbed. Texas banded geckos eat termites and other insects, and their scat is like that of other lizards. It is usually a long brown or black pellet with a white cap of uric acid.

HOW TO SPOT

Scat Size: 0.5 inches (1.3 cm) long

Range: Southern United States and northeastern Mexico

Habitat: Deserts and dry, rocky canyons

Diet: Spiders, crickets, moths, and termites

TEXAS HORNED LIZARD

(PHRYNOSOMA CORNUTUM)

Texas horned lizards, also called horny toads, live in deserts, grasslands, and scrublands. These lizards have many predators and defend themselves by inflating their usually flat bodies, making their horns appear larger. And they spray foul-smelling blood from their eyes to scare predators away. They also have rough, spiky skin that helps them blend into their environments. Texas horned lizards eat ants and other insects and invertebrates. Their droppings are small, tubular, and dark with a dollop of uric acid at the tip. When crushed, the feces is composed mostly of ants.

HOW TO SPOT

Scat Size: About 1 inch (2.5 cm) long

Range: South-central and southwestern United States and northern Mexico

Habitat: Arid and semiarid areas with little vegetation

Diet: Ants and other insects

GLOSSARY

alpine
Having to do with mountains.

archipelago
A group of islands.

boreal
Located in northern regions.

camoflauge
An animal's coloring or marks that help it blend in with its surroundings.

carrion
The flesh and bones of a dead animal that becomes a meal for other animals.

defecate
To expel feces.

dewclaw
A toe on the back of the foot on many animals. It grows higher on the leg than other toes.

forb
An herb plant that flowers but is not woody or a grass.

gait
A manner or method of walking or moving.

hallux
A large toe on a bird that faces backward.

plumage
Bird feathers.

protein
A substance that bodies need to function.

regurgitate
To bring swallowed food back up to the mouth.

scat
Feces or poop.

talon
A claw on the feet of birds of prey.

terrestrial
Of or relating to land.

undulate
To rise and fall smoothly.

ungulate
A mammal with hooves.

TO LEARN MORE

FURTHER READINGS

Dendy, Leslie. *Tracks, Scats and Signs*. Cooper Square Publishing, 1996.

Giménez de Ory, Beatriz. *Poop!* Barefoot Books, 2020.

Halfpenny, James. *Scats and Tracks of North America: A Field Guide to the Signs of Nearly 150 Wildlife Species*. Falcon Guides, 2019.

Levine, Lynn. *Mammal Tracks and Scat: Life-Size Pocket Guide*. Heartwood Press, 2014.

ONLINE RESOURCES

Booklinks
NONFICTION NETWORK
FREE! ONLINE NONFICTION RESOURCES

To learn more about tracks or scat, please visit **abdobooklinks.com** or scan this QR code. These links are routinely monitored and updated to provide the most current information available.

PHOTO CREDITS

Cover Images: Amelia Martin/Shutterstock, front (bear scat); Charles T. Peden/Shutterstock, front (tortoise tracks); Dan Kosmayer/Shutterstock, front (bear tracks); Doug McLean/Shutterstock, front (moose tracks); Jerry Lin/Shutterstock, front (dog track); PixieBitz/Shutterstock, front (cat tracks); saodaeng/Shutterstock, front (gecko scat); schankz/Shutterstock, front (bird scat); ShadowMaster/Shutterstock, front (grouse scat); Tim Burrett/Shutterstock, front (bird tracks); Vistacoloris/Shutterstock, front (swan scat); weha/Shutterstock, front (goat scat); Max Allen/Shutterstock, back (scat); Robert Hoetink/Shutterstock, back (bird tracks)
Interior Images: Chamois huntress/Shutterstock, 1 (top left), 5 (top center), 24 (right); Arterra/Universal Images Group/Getty, 1 (bottom left), 60 (top); Verushka/Shutterstock, 1 (top center), 71 (top); Orange Grove/Shutterstock, 1 (bottom center), 4 (bottom left), 29 (top); K Steve Cope/Shutterstock, 1 (top right), 19 (left), 28 (bottom); Matyas Rehak/Shutterstock, 1 (bottom right), 78 (right); Ipoel Jangan Asem/Shutterstock, 4 (center), 84 (left); Tyler Olson/Shutterstock, 4 (top right), 45 (right); Monikabaumbach/Dreamstime.com, 4 (bottom right), 12 (right); KathrynSK/Getty, 5 (top left), 99 (bottom); Nature Photographers Ltd/Alamy Stock Photo, 5 (top right), 95 (right); Ryley MacKinnon/Shutterstock, 5 (center left), 26 (left), 112 (bottom left); Randy Bjorklund/Shutterstock, 5 (center right), 88 (right), 100 (bottom); Danae Abreu/Shutterstock, 5 (bottom left), 63 (top); pran/Shutterstock, 5 (bottom right), 32 (right); Ernie Cooper/Shutterstock, 8 (left); Tosh Child/Shutterstock, 8 (right); Nick Olson/Shutterstock, 9; Chase D'animulls/Shutterstock, 10 (left), 21 (right); K Alina/Shutterstock, 10 (right); EnelGammie/Shutterstock, 11 (top); Rosamar/Shutterstock, 11 (bottom); Elliotte Rusty Harold/Shutterstock, 12 (left); Gilbert S. Grant/Shutterstock, 13 (top); Pete Holley/Shutterstock, 13 (bottom); Chris Hill/Shutterstock, 14 (left), 16 (right); Melani Wright/Dreamstime.com, 14 (right); David G Hayes/Shutterstock, 15 (left); Dennis W Donohue/Shutterstock, 15 (right); Sergey 402/Shutterstock, 16 (left); matunka/Shutterstock, 17 (left); Robert Adami/Shutterstock, 17 (right); GreenEarth Photography/Shutterstock, 18 (left), 112 (center); Aaron J Hill/Shutterstock, 18 (right); Aksenova Natalya/Shutterstock, 19 (right); AndreAnita/Shutterstock, 20 (left), 47 (right); Shyjo/Shutterstock, 20 (right); James Neal/Dreamstime.com, 21 (left); Ashley Keller/Shutterstock, 22 (left); Danita Delimont/Shutterstock, 22 (right), 70 (right), 85 (bottom); Karel Bock/Shutterstock, 23 (left), 30 (top), 83 (bottom); ressormat/Shutterstock, 23 (right); Sophia Granchinho/Shutterstock, 24 (left); Wirestock Creators/Shutterstock, 25 (left); gatito33/Getty, 25 (right); critterbiz/Shutterstock, 26 (right); Kerry Hargrove/Shutterstock, 27 (top), 64 (bottom); Arterra Picture Library/Alamy Stock Photo, 27 (bottom); Eric Isselee/Shutterstock, 28 (top), 34 (right); Jim Cumming/Shutterstock, 29 (top), 50 (right), 62 (right), 87 (bottom); Jomegat/Wikimedia Commons, 30 (bottom); gillmar/Shutterstock, 31 (left); Mimka/Shutterstock, 31 (right); Dora Zett/Shutterstock, 32 (left); Deb Anderson/Shutterstock, 33 (left); Agami Photo Agency/Shutterstock, 33 (right); Gerard Soury/Getty, 34 (left); OlesyaNickolaeva/Shutterstock, 35 (left); Estelle R/Shutterstock, 35 (right); Sergey Uryadnikov/Shutterstock, 36 (left); Jens Otte/Shutterstock, 36 (right); Agnieszka Bacal/Shutterstock, 37 (left), 51 (right), 81 (top); Jeff Holcombe/Shutterstock, 37 (right); NancyS/Shutterstock, 38 (left); Eric Dale/Shutterstock, 38 (right); Lehmanphotos/Dreamstime.com, 39 (left); George D. Lepp/Getty, 39 (right); Mi St/Shutterstock, 40 (top); Haggardous50000/Shutterstock, 40 (bottom); Marcelo Morena/Shutterstock, 41 (top); Sari ONeal/Shutterstock, 41 (bottom); Mia & Steve Mestdagh/Flickr, 42 (left); Tom Reichner/Shutterstock, 42 (right); AB Photographie/Shutterstock, 43 (left); Okyela/Shutterstock, 43 (right); Saad315/Shutterstock, 44 (left); Chris

and Tilde Stuart/FLPA/Minden Pictures, 44 (right); Vaclav Sebek/Shutterstock, 45 (left); Zeljkokcanmore/Dreamstime.com, 46 (left); Anan Kaewkhammul/Shutterstock, 46 (right); Kristine Rad/Shutterstock, 47 (left); Jukka Jantunen/Shutterstock, 48 (left), 50 (left), 96 (right); Rabbitti/Shutterstock, 48 (right), 82 (top); Cultura RM Exclusive/Jouko van der Kruijssen/Getty, 49 (left); LukasPich/Shutterstock, 49 (right); P. Qvist/Shutterstock, 51 (left); Lumppini/Shutterstock, 52 (left); WildMedia/Shutterstock, 52 (right); oksmit/Shutterstock, 53 (left); Alice Cahill/Getty, 53 (right); Michael Rolands/Shutterstock, 54 (top); reptiles4all/Shutterstock, 54 (bottom); John Sullivan/Alamy Stock Photo, 55; Jon G. Fuller/VWPics/Newscom, 56 (left); Matthijs Kuijpers/Dreamstime.com, 56 (right); Mark_Kostich/Shutterstock, 57 (top); Chantelle Bosch/Shutterstock, 57 (bottom); Ale_Koziura/Shutterstock, 58 (left); Harry Collins Photography/Shutterstock, 58 (right); Feng Yu/Shutterstock, 59; Jay Gao/Shutterstock, 60 (bottom); drferry/Getty, 61; Andria C Johnston/Shutterstock, 62 (left); Paul Jones Jr/Shutterstock, 63 (bottom); Derek Wallentinsen/Dreamstime.com, 64 (top); Jose A. Bernat Bacete/Getty, 65; Harold Stiver/Shutterstock, 66 (left); Peggy Hazelwood/Shutterstock, 66 (right), 92 (bottom); Aria_RJWarren/Shutterstock, 67 (top); Vistacoloris/Shutterstock, 67 (bottom); Phil Champion/Geograph, 68 (left); Charles T. Peden/Shutterstock, 68 (right), 105 (bottom), 107 (top); Lee H. Rentz/NHPA/Photoshot/Newscom, 69 (top); Frank Fichtmueller/Shutterstock, 69 (bottom); HelenL100/Getty, 70 (left); J. Andrew Patronik/Shutterstock, 71 (bottom); Jay Ondreicka/Shutterstock, 72 (top), 75 (left), 112 (top right); LifeLoveLiberty/Shutterstock, 72 (bottom); Max Allen/Shutterstock, 73 (top); Reimar/Shutterstock, 73 (bottom), 84 (right); Milan Zygmunt/Shutterstock, 74 (bottom); Holger Kirk/Shutterstock, 75 (right); Kokoulina/Shutterstock, 76 (top); Daniel Avram/Shutterstock, 76 (bottom); James Weliver/USFWS/Flickr, 77 (left); Tomas Kulaja/Shutterstock, 77 (right); Dirk M. de Boer/Shutterstock, 78 (left); Michele Aldeghi/Shutterstock, 79 (left); Jody./Shutterstock, 79 (right); David Pimborough/Shutterstock, 80 (left); Martin Rudlof Photography/Shutterstock, 80 (right); suphanat/Shutterstock, 81 (bottom); Daria Soroka/Shutterstock, 82 (bottom); Stefan Winberg/Shutterstock, 83 (top); josera/Shutterstock, 85 (top); stockphotofan1/Shutterstock, 86 (left); Bruce MacQueen/Shutterstock, 86 (right); Miholcea Zsolt/Shutterstock, 87 (top); Henryk Sadura/Shutterstock, 88 (left); Photok.dk/Shutterstock, 89 (left); Alexander_Safonov/Shutterstock, 89 (right); sbonk/Getty, 90 (left); James Hager/robertharding/Newscom, 90 (right); Szasz-Fabian Jozsef/Shutterstock, 91 (top); sophiecat/Shutterstock, 91 (bottom); Teresa Kopec/Getty, 92 (top); Angie Westre/Dreamstime.com, 92 (bottom); Seney Natural History Association/Wikimedia Commons, 93 (top); vadimserebrenikov/Shutterstock, 94 (left); Eivor Kuchta/Shutterstock, 94 (right); carlosbezz/Getty, 95 (left); Jason Ondreicka/Dreamstime.com, 96 (left); Tim Malek/Shutterstock, 97 (left); aydngvn/Shutterstock, 97 (right), 112 (bottom right); Max Allen/Alamy Stock Photo, 98 (left); nvphoto/Shutterstock, 98 (right); Matt Knoth/Shutterstock, 99 (top); PRE/Wikimedia Commons, 100 (top); GSPhotography/Shutterstock, 101 (top); Lisa Wright/Animal Diversity Web, 101 (bottom); Holly Kuchera/Shutterstock, 102 (top); Michael Sean OLeary/Shutterstock, 102 (bottom); moosehenderson/Shutterstock, 103 (top); Adam Cegledi/Shutterstock, 103 (bottom); phichak/Shutterstock, 104 (top); Alexandr Junek Imaging/Shutterstock, 104 (bottom); Gchapel/Shutterstock, 105 (top); saodaeng/Shutterstock, 106 (top), 112 (top left); David A. Northcott/Getty, 106 (bottom); Matt Jeppson/Shutterstock, 107 (bottom)

ABDOBOOKS.COM
Published by Abdo Reference, a division of ABDO, PO Box 398166, Minneapolis, Minnesota 55439.
Printed in China

102023
012024

THIS BOOK CONTAINS
RECYCLED MATERIALS

Editor: Leah Kaminski
Series Designer: Colleen McLaren

Library of Congress Control Number: 2023939625
Publisher's Cataloging-in-Publication Data
Names: Russo, Kristin J. , author.
Title: Tracks and scat / by Kristin J. Russo
Description: Minneapolis, Minnesota : Abdo Reference, 2024 | Series: North American field guides | Includes online resources.
Identifiers: ISBN 9781098293116 (lib. bdg.) | ISBN 9798384911050 (ebook)
Subjects: LCSH: Animal tracks--Juvenile literature. | Scat (Feces)--Juvenile literature. | Animal signs--Juvenile literature. | Animal behavior--Juvenile literature. | Encyclopedias and dictionaries--Juvenile literature.
Classification: DDC 591.5--dc23